U.S. CIVIL
AIRCRAFT SERIES
VOLUME 3

This work is dedicated to the preservation and perpetuation of a fond memory for the men and the planes that made a future for our air industry. And, to help kindle a knowledge and awareness within us for our debt of gratitude we owe to the past. . . . .

U.S. CIVIL AIRCRAFT SERIES

VOLUME 3

(ATC 201 - ATC 300)

Joseph P. Juptner

TAB *AERO*

Division of McGraw-Hill, Inc.

Blue Ridge Summit, PA 17294-0850

ACKNOWLEDGEMENT

Any historian soon learns that in the process of digging for obscure facts and information, he must often times rely on the help of numerous people, unselfish and generous people, many who were close to or actually participated in the various incidents or events that make up this segment of history recorded here and have been willing to give of their time and knowledge in behalf of this work. To these wonderful people I am greatly indebted and I feel a heartwarming gratitude; it is only fitting then, that I proclaim their identity here in appreciation.

My thanks to Alfred V. Verville; Chas. W. Meyers; Eddie Martin; Earl C. Reed; to A. W. French of Pan American World Airways; Gerald Deneau of Cessna Aircraft Co.; Ray Silvius of Western Air Lines; Harvey Lippincott of Pratt & Whitney Div.; Philip S. Hopkins, S. Paul Johnston and staff at Smithsonian Institution, National Air Museum; Beech Aircraft Corp.; American Air Lines; United Air Lines; Convair Div.; Gordon S. Williams of the Boeing Co.; Ken Molson of National Aviation Museum of Canada; Fairchild-Stratos; and the following group of dedicated enthusiasts and historians; Stephen J. Hudek; Roy Oberg; A. Jack McRae; Wm. T. Larkins; John W. Underwood; Marion Havelaar; Roger Besecker; Truman C. Weaver; Joe Christy; Dave Jameson; Ralph Nortell; Louis M. Lowry; Richard Sanders Allen; John R. Ellis; Burton Kemp; Douglas E. Anderson; James W. Bott; Chas. F. Schultz; Peter M. Bowers; and Gerald H. Balzer.

FOREWORD

Year 1929 as we remember it was quite an exciting and bewildering year in aviation progress; it seemed that each monthly issue of the airplane magazines announced the formation of from 3 to 6 new "airplane factories," there was nearly a 200% increase in licensed pilots, domestic air-mail poundage was more than doubled, scheduled air-line operations increased from 27,000 to 100,000 miles daily, 166 aircraft "types" had been approved that year by the over-worked Dept. of Commerce, 19 new engines received their certificates of approval, more and more people were taking to aviation, and every facet of the aviation trade was riding high, wide, and handsome.

Then, like a sudden storm lashing out of nowhere and without perceptible warning, came circumstances that caused the onset of our great economic depression. Needless to say, everyone was rocked back on their heels in bewilderment, without a full realization of just what had happened; everybody was at first skeptical, somewhat worried, and had no immediate plans on how to cope with this situation. Soon to add to this dilema came cancellations of airplane orders which in turn forced manufacturers to cancel orders for engines, propellers, and raw materials; indeed, this had a sobering effect on all manufacturers and operators in the business, and they were taking time to think. Now with all the froth and gayety just about gone, the aviation industry was forced to take stock of itself and a good hard look at its predicament; by and large it soon came to realize that it had better become a more sensible and more stable member of the industrial family if it was to survive. This was to take much effort and quite a bit of heart-break but it was something that was inevitable and had to be done.

Despite the severe blow that had shaken it to its very roots, the year 1929 was a very interesting period in the annals of airplane development. It was also a time when most of the pioneers of aviation, around whom the romance and adventure of early airplane development was woven, were beginning to share the stage with many bright-eyed young men coming up with a new out-look, who perhaps a decade from now would surely be the giants in charge of a vast and comparatively healthy industry. It was a time too when the aura of romance and adventure was binding itself into an era of a sobering industry that was forced to think more carefully and be more practical. There were lean years ahead, many operated on that proverbial shoestring and many were finally forced to close their doors but progress in airplane development pushed forward just the same, if not more so. It has often been said in jest that aviation people are at their best when hungry; the outstanding developments brought forth in the next few years would almost lead one to believe it.

Joseph P. Juptner

TERMS

To make for better understanding of the various information contained herein, we should clarify a few points that might be in question. At the heading of each new chapter, the bracketed numerals under the ATC number, denote the first date of certification; any amendments made are noted in the text. Unless otherwise noted, the title photo of each chapter is an example of the model that bears that particular certificate number; any variants from this particular model, such as prototypes and special modifications, are identified. Normally accepted abbreviations and symbols are used in the listing of specifications and performance data. Unless otherwise noted, all maximum speed, cruising speed, and landing speed figures are based on sea level tests; this method of performance testing was largely the custom during this early period. Rate of climb figures are for first minute at sea level, and the altitude ceiling given is the so-called service ceiling. Cruising range in miles or hours duration is based on the engine's cruising r.p.m., but even at that, the range given must be considered as an average because of pilot's various throttle habits.

At the ending of each chapter, we show a listing of registered aircraft of a similar type; most of the listings show the complete production run of a particular type and this information we feel will be valuable to historians, photographers, and collectors in making correct identification of a certain aircraft by its registration number.

In each volume there are separate discussions on 100 certificated airplanes and we refer to these discussions as chapters, though they are not labeled as such; at the end of each chapter there is reference made to see the chapter for an ATC number pertaining to the next development of a certain type. As each volume contains discussions on 100 aircraft, it should be rather easy to pin-point the volume number for a chapter of discussion that would be numbered as A.T.C. #93, or perhaps A.T.C. #176, as an example. The use of such terms as "prop", "prop spinner", and "type", are normally used among aviation people and should present no difficulty in interpreting the meaning.

TABLE OF CONTENTS

A.T.C. #201
(8-16-29)
MONO "MONOCOACH"

Fig. 1. Monocoach model 201 with Wright J5 engine; shown here at factory.

With "Monocoupe" production moving along at a good pace, and the popularity of the pixie-like cabin monoplane at a high pitch, Mono Aircraft was grooming the four-place "Monocoach" as a big-sister addition to the swelling Mono line. Seating 4 people with ample room and good comfort, the new model "201" was a high-wing cabin monoplane of typical family resemblance that was powered with the 9 cyl. Wright J5 engine of 220 h.p.; this was a combination that had plenty of reserve power to offer a rather high performance. The development of the 4 place cabin monoplane of this particular type, was brought about by requests from numerous private-owners for a larger family-type airplane; most would-be owners specified a cabin craft with seating for no more than four but with plenty of practical comfort, a capacity for extra payload, and above all a rather high performance throughout the whole range of operation to cope with non-scheduled flying under any of varied conditions. As a high-performance family-type airplane or efficient air-taxi, the "Monocoach" 201 with

its pleasant proportion and sturdy construction had no peer; ideal for men of business, it was also designed as a money-maker for the flying service operator. The past performance of this craft, as since recorded, will now bear out that the big-sister version of the little "Monocoupe" was indeed a sturdy breed of airplane that had taken its place proudly in the panorama of aviation development as one of the best airplanes of this type.

Developed in the latter part of 1928 from designs by Clayton Folkerts, the prototype "Monocoach" (X-8900) was first powered for tests with the newly developed Velie L-9 engine of 185 h.p.; the L-9 was a 9 cyl. air-cooled "radial" type engine of quite normal design but much development was yet due in order to iron out some of the problems that had cropped up from time to time. Anxious to offer the "Monocoach" to the favorable market that was present, the Wright J5 engine was selected as an alternate power-plant until such a time when the 9 cyl. Velie L-9 could be pronounced a success; as it turned out, the L-9 development did not warrant the expense involved and further experimentation was cancelled. Incidently, the "Monocoach" was first planned as a larger 6-seater but the clamor for a high-performance 4-seater ruled this project out. As promotion for the newly-designed "201", the craft was entered in the 1929 National Air Races held at Cleveland, Ohio. Flown in the Philadelphia to Cleveland Air Derby by "Ike" Stewart the "Monocoach" was bested only by the fast-flying Alexander "Bullet". Also on demonstration during air-race week, the verve and personality of the new "201" was quite impressive. A promotion tour throughout California was flown by Don A. Luscombe, and a nation-wide tour sponsored by the Elks Club, brought the "Monocoach" to public attention around the country.

The type certificate number for the "Monocoach" model 201 as powered with the 220 h.p. Wright J5 engine was issued 8-16-29, at a time when nearly 500 assorted airplanes had already been built by Mono Aircraft and the firm was happily celebrating its third anniversary since the development of their first product, the Central States "Monocoupe", in August of 1926. The "Monocoach" 201 was manufactured by the Mono Aircraft, Inc. at Moline, Illinois and some 14 or more examples of this model were built. Don A. Luscombe was the president; and Clayton Folkerts was the chief engineer; both talented and gifted men that had a knack of conceiving some of the finest airplanes this country has ever known.

Listed below are specifications and performance data for the "Monocoach" model 201 as powered with the 220 h.p. Wright "Whirlwind" J5 engine; length overall 26'8"; hite overall 8'7"; wing span 39'0"; wing chord 75"; total wing area 230 sq.ft.; airfoil section USA-35B; wt. empty 1919; useful load 1173; payload with 63 gal. fuel was 578 lbs.; gross wt. 3092 lbs.; max. speed 128; cruising speed 110; landing speed 50; climb 900 ft. first min. at sea level; ceiling 18,000 ft.; gas cap. 63 gal.; oil cap. 6 gal.; cruising range was 5 hours or 550 miles; price at the factory field was $7950.; for a time the "201" was the cheapest cabin airplane of this type; the "Monocoach" with Velie L-9, had it been produced in number, was to sell for $6500.

The fuselage framework was built up of

welded chrome-moly and 1025 steel tubing, faired to shape with wood formers and fairing strips and fabric covered. The interior was tastefully upholstered in mohair fabric and shatter-proof glass was installed throughout; there was a sky-light over the pilot's area and windows in the doors slid up and down for ventilation. There was a large cabin door and a convenient step on each side for ease of entry; front seats were individual with fold-down backs and the rear seat was of the bench type with an easily accessible baggage compartment behind it. The semi-cantilever wing framework was built up of solid spruce spars that were routed to an I-beam section; truss-type wing ribs were of spruce and basswood, the leading edge was covered with dural sheet to the front spar and the completed framework was fabric covered. The 2 gravity-feed fuel tanks were mounted in the root end of each wing flanking the fuselage, and the wing was braced by two parallel struts on each side; the landing gear employed spring-draulic shock absorbing struts and was incorporated into a sturdy truss with the wing-bracing struts. The wide tread landing gear was fitted with 30x5 Bendix wheels and brakes, and a full-swivel tail wheel was provided for ease in ground maneuvering. The fabric covered tail-group was built up of welded chrome-moly and 1025 steel tubing; the fin was ground adjustable and the horizontal stabilizer was adjustable in flight. Optional color schemes were available, but most examples of the "201" were a shiny black with bright orange-yellow. A metal propeller, inertia-type engine starter, navigation lights, and dual stick controls were standard equipment. For the next development in the "Monocoach" series, see the chapter for ATC # 275 in this volume; another interesting development in the "Mono" line was the lovable "Monoprep" trainer which is discussed in the chapter for ATC # 218 in this volume.

Listed below are "Monocoach" model 201 entries that were gleaned from various registration records; as far as can be determined, these were the only examples built:

X-8900 ; Monocoach 201 (# 5000) Velie L-9
C-8916 ;   „   (# 5001) Wright J5
C-8953 ;   „   (# 5002)   „
C-8960 ;   „   (# 5003)   „
C-8969 ;   „   (# 5004)   „
C-8971 ;   „   (# 5005)   „
NC-8976 ;   „   (# 5006)   „

Fig. 2. Monocoach was popular family-type airplane.

C-8978 ;	,,	(# 5007)	,,
C-8981 ;	,,	(# 5008)	,,
NC-8988 ;	,,	(# 5009)	,,
NC-8996 ;	,,	(# 5010)	,,
NC-100K ;	,,	(# 5011)	,,
NC-114K ;	,,	(# 5012)	,,
NC-116K ;	,,	(# 5013)	,,
NC-125K ;	,,	(# 5014)	,,

Serial # 5000 was prototype for Monocoach series; serial # 5001 was prototype for 201 series; serial # 5002 thru # 5007 on Group 2 approval numbered 2-109; serial # 5008 thru # 5014 on ATC # 201; serial # 5011 and # 5012 later converted to model 275 with Wright J6-7-225 engine.

A.T.C. #202
(8-16-29)
GOLDEN EAGLE, "CHIEF"

Fig. 3. Sporty Golden Eagle "Chief" with 90 H.P. Le Blond engine.

The Golden Eagle "Chief" was a tidy-looking airplane of normal lines that had been especially designed for the private-pilot; though advertised and recommended as a craft suitable for pilot-training or an economical air-taxi for the business-man pilot, it nevertheless turned out to be just a sport-type airplane well suited to flying for the fun of it. Many, many airplanes, though capable flying machines, were bally-hooed beyond all reason as craft able to serve a multitude of useful purposes, but a good many were actually very little more than airplanes that were just fun to fly around in and "build up time" with. The Golden Eagle "Chief" as shown here, was a parasol-type open cockpit monoplane that seated two in tandem and was powered with the 7 cyl. LeBlond 7-D engine of 90 h.p. This well-rounded beauty had plenty of performance to offer and most of its flight characteristics were quite normal, except for "bad spinning habits" that were later over-come by a slight redesign of the tail surfaces. Eddie Martin, a veteran air-mail pilot who performed many of the early test-flights in the Golden Eagle "Chief", had to "bail out"

of one and take to his 'chute because "it wound up tight" and wouldn't recover to normal attitude. The bulbuous N.A.C.A. low-drag engine cowling and buxom fuselage profile were suspected of blanking off the tail-group and therefore the lack of response; records show that Lee Flanigan and Jack Rahn also had to bail out of the "Golden Eagle" before the trouble was finally corrected. In its final form, the "Chief" was a pretty fair ship and no doubt many more would have been seen around the country, if the bottom hadn't fallen out of the aircraft industry at this particular time.

The "Golden Eagle" design in its original form was conceived somewhat differently by Mark M. Campbell, a talented and versatile man who had been wing-walker, parachute jumper, airplane mechanic, pilot, and aircraft designer since 1915. The first 3 "Golden Eagle" were 1-seated sport monoplanes with 3 cyl. and 6 cyl. Anzani engines, and one with a LeBlond 60; a 2-seated craft followed that was powered with both the Velie and LeBlond engines. This 2-seated version was of the type

Fig. 4. Early Golden Eagle with Velie engine; designed by Mark Campbell

Fig. 5. Deep fairing and engine cowling made "Chief" a handsome ship but caused early problems.

used by 18 year-old "Bobbie" Trout to set an endurance record (non-refueled) for woman pilots of over 17 hours of continuous flight. Mark M. Campbell teamed up with R. O. Bone, a retired business-man, to form a company called R. O. Bone & Associates in Inglewood, Calif.; both attended the 1928 National Air Races held in Los Angeles to show their new craft to the flying public. Shortly after "Bobbie" Trout's endurance flights, there was some misunderstanding on plans for the future between Bone and Campbell, so Campbell decided to leave and design chores were turned over to F. M. Smith. Smith then redesigned the "Golden Eagle" to incorporate a N.A.C.A. type low-drag engine cowling, plus an increase in wing area and an extensive re-fairing of the fuse-lage profile; it was these changes that caused an unhappy balance in the aerodynamic character of the basic "Golden Eagle" design. After Campbell had left, the company was re-organized in early 1929 as the Golden Eagle Aircraft Corp. of Inglewood, Calif. A Group 2 approval numbered 2-107 was issued 8-12-29 for serial # 803-804-805; an approved type certificate number was issued 8-16-29 for serial # 806 thru # 810. As far as records show, some 10 examples of the Golden Eagle "Chief" were built in the plant at Inglewood, Calif.; R. O. Bone was the president and F. M. Smith was secretary and chief engineer. Hoping to secure refinancing, the company moved its operations to a leased hangar on the munici-pal airport in Port Columbus, Ohio but several circumstances prevented the re-located operation to blossom forth as planned. In 1930, veteran pilot "Joe" Mackey and several associates bought up the remnants of the Golden Eagle company and moved it to Lancaster, Ohio where they had planned to build the "Chief" on a spare-time basis. It has

been once reported that Mackey and associ-ates built 27 of the G.E. "Chief" but it is very doubtful if that many were ever built. By 1931, Joe Mackey had moved his operations to Findlay, Ohio where he formed the famous "Linco Flying Aces" stunt-flying team and manufacture of the "Chief" had since been suspended. It is gratifying to know that at least one example of the Golden Eagle "Chief" survived and yet remains to provide a Spokane airman many hours of flying pleasure.

Listed below are specifications and per-formance data for the Golden Eagle "Chief" as powered with the 90 h.p. LeBlond 7-D engine; length 21'0"; hite 7'6"; wing span 30'6"; wing chord 66"; total wing area 165 sq.ft.; airfoil Clark Y; wt. empty 966; useful load 513; payload with 25 gal. fuel was 174 lbs.; gross wt. 1479 lbs.; max. speed 120; cruising speed 98; landing speed 38; climb 970 ft. first min. at sea level; ceiling 12,000 ft.; gas cap. 25 gal.; oil cap. 3 gal.; cruising range at 6.5 gal. per hour was 375 miles; price at the factory was $3950, later lowered to $3650. and $3390. and finally $2990.; a Wright-Gipsy version of the "Chief" was offered for $3750.

The fuselage framework was built up largely of welded 1025 steel tubing, with welded chrome-moly steel tubing at all highly stressed points; the framework was heavily faired to shape with wood formers and fairing strips and fabric covered. The wing framework was built up of laminated spruce spars that were routed to an I-beam section; wing ribs were built up of spruce and plywood in truss-type form, the leading edges were covered in plywood, and the completed framework was covered with fabric. There was one fuel tank in the root end of each wing-half; normal tanks were of 12.5 gal. capacity but two 20 gal. tanks were available for a total capacity of 40 gal. Wing bracing struts varied from the vee-type on some early examples, to N-type struts on later examples, with slight changes in the center-section cabane. The split-axle landing gear was built up of welded chrome-moly steel tubing and used rubber shock-cord to absorb the bumps; wheels were 26x4 and brakes were available. Cockpits were deep and roomy and there was a large door in left side of fuselage for entry to front cockpit; there was no allowance for baggage. The 7 cyl. LeBlond engine was cowled-in tightly with a N.A.C.A. type low-drag cowling that straightened out the air-flow around the engine cylinders; the gain in max. speed was better than 5 m.p.h. Streamlined wheel fairings, commonly called "wheel pants", were also available to add to the top speed. The LeBlond 7-D engine of 90 h.p. was identical to the 5-D of 65 h.p. except for addition of 2 cylinders around a different crank-case; 90 per cent of the parts were interchangeable. In 1930, one example of the "Golden Eagle" was powered with the 5 cyl. Kinner K5 engine; this was the "Junior Pursuit" version which was to sell for $3850.

Listed below are "Golden Eagle" entries of all models that were gleaned from various registration records:

X-7383:	R. O. Bone	(# 701)	
			6 cyl. Anzani.
C-10071:	,,	(# 702)	Velie 55
-816N;	,,	(# 703)	,,
X-98E;	,,	(# 704)	,,
X-3335:	Chief	(# 801)	LeBlond 90
-522;	,,	(# 802)	,,
X-569K;	,,	(# 803)	,,
;	,,	(# 804)	,,
C-245M;	,,	(# 805)	,,
NC-524M;	,,	(# 806)	,,
;	,,	(# 807)	,,
NC-68N;	,,	(# 808)	,,
NC-56W;	,,	(# 809)	,,
-10057;	,,	(# 810)	,,

Serial # 701-702-703-704 also as # 1-2-3-4, were mfgd. by R. O. Bone & Assoc.; registration numbers for serial # 801-802 unverified; serial # 803-804-805 on Group 2 approval numbered 2-107; serial # 806 thru # 810 on ATC # 202; serial # 804 operated in Alaska, registration number unknown; registration number unknown for serial # 807 which might have been CF-AKB; X-254V was "Jr. Pursuit", serial # 2002; there may have been a serial # 2001, but none was listed.

Fig. 6. Golden Eagle offered year-round sport flying.

Fig. 7. Veteran Golden Eagle "Chief" still flying in 1963.

A.T.C. #203
(8-23-29)
AEROMARINE-KLEMM, AKL-26

Fig. 8. Aeromarine-Klemm model AKL-26 with 60 H.P. Le Blond engine.

Though quite suitable as a private-owner pleasure craft that could operate on a minimum budget, and certainly a sensible arrangement for a training airplane that would pay off in profits, the 40 horsepower AKL-25 was not met with any surging enthusiasm in this country; a seeming lack of favor perhaps because of the average American's exposure to aircraft of much higher horsepower. Any 2-seated craft that had less than 55-66 h.p. was considered underpowered and the Salmson-powered Aeromarine-Klemm was practically scoffed at as a "flying flivver". Born of a glider design in the first place, the little AKL actually was a flying flivver in a sense, but the performance and utility dished out by this little combination was quite phenomenal in direct proportion to its available power. In spite of this, the design-engineers at Aeromarine-Klemm had to give in to the inevitable, with some reluctance, and cater to the standard policy that governed the trend in this country; this trend being the constant increases in engine horsepower to gain more of the performance

that would be attractive to buyers of aircraft, whether for pleasure or profitable gain.

The Aeromarine-Klemm model AKL-26 was a low-winged open cockpit monoplane seating two in tandem; the basic configuration was typical to previous models except for the increase in horsepower which was supplied by the 5 cyl. LeBlond 60 (5-D) of 65 h.p. at 1950 r.p.m. First tested in a prototype version as the AKL-60, a noticeable improvement in performance was naturally gained by the increase in power; inherent stability and good flight characteristics were retained, and all in all, the new model had shown good promise of being well accepted. Several flying-schools added the higher-powered AKL-26 to their flight line, and a few were operated strictly as pleasure craft. Starting off with a lively spurt in sales, the AKL-26 was well on its way to being produced in large numbers, but the economic sag in this country curtailed its full potential drastically. The model AKL-26 as powered with the Le-Blond engine of 65 h.p., received its type cer-

Fig. 9. AKL-26 offered low-cost sport flying.

tificate number on 8-23-29 for landplane and seaplane, and some 5 or more examples of this particular model were manufactured by the Aeromarine-Klemm Corp. at Keyport, New Jersey. Inglis M. Uppercu was the president, and the well-known Vincent J. Burnelli was V.P. in charge of engineering. The Aeromarine-Klemm Corp. was organized to manufacture the German-designed Klemm monoplane in June of 1928; although this firm was a relatively new organization, Uppercu and "Aeromarine" had been active in the aviation industry since 1908.

Listed below are specifications and performance data for the Aeromarine-Klemm model AKL-26 as powered with the 65 h.p. LeBlond engine; length overall 23'6"; hite overall 7'0"; wing span 40'2"; wing chord at root 79"; total wing area 194.5 sq.ft.; airfoil Goettingen 387 modified; wt. empty (landplane) 940 lbs.; useful load 500; payload with 15 gal. fuel was 230 lbs.; gross wt. 1440 lbs.; max. speed 93; cruising speed 80; landing speed 38; climb 450 ft. first min. at sea level; ceiling 10.000 ft.; gas cap. 15 gal.; oil cap. 2 gal.; cruising range at 4.5 gal. per hour was 245 miles; price at the factory was $3500. The following figures are for twin-float seaplane; wt. empty 1090; useful load 500; payload with 15 gal. fuel was 230 lbs.; gross wt. 1590 lbs.; max. speed 85; cruising speed 75; landing speed 40; climb 375 ft. first min. at sea level; ceiling 8500 ft.; cruising range 225 miles; price at factory $4500, with Edo H twin-float gear.

The fuselage framework was built up of spruce longerons, with plywood bulkheads and formers, covered with plywood veneer to form a semi-monocoque shell; a gravity-

Fig. 10. Prototype AKL-26 in early tests.

Fig. 11. AKL-26 offered added pleasure as seaplane.

feed fuel tank was mounted high in the fuselage, just ahead of the front cockpit. The
cantilever wing framework was built up of
spruce and plywood box-type spar beams and
plywood wing ribs; the completed framework was covered with plywood veneer. The
split-axle landing gear of 85 inch tread used
oleo-spring shock absorbing struts; wheels
were 24x4 or 20x9. Fixed surfaces of the tail-
group were built up of spruce spars and
former-ribs, covered in plywood veneer;
movable surfaces were built up of spruce
spars and former-ribs, covered in fabric. The
vertical fin and horizontal stabilizer were
adjustable only on the ground. A baggage
compartment with allowance for 55 lbs. was
located in the turtle-back section of the fuselage; baggage allowance included anchor and
rope, when carried in seaplane. The next
development in the Aeromarine-Klemm
monoplane was the model AKL-26A as discussed in the chapter for ATC # 204 in this
volume.

Listed below are AKL-60 and AKL-26
entries that were gleaned from registration
records:

NC-120H;	AKL-60	(# 2—14)	LeBlond 60
NC-122H;	„	(# 2—16)	„
NC-123H;	„	(# 2—17)	„
NC-52K ;	AKL-26	(# 2—29)	„
-100M;	„	(# 2—30)	„
NC-101M;	„	(# 2—32)	„
NC-102M;	„	(# 2—33)	„
NC-753N;	„	(# 86)	„

Serial # 2-16 and # 2-17 first had Salmson
40 as AKL-25A, modified to AKL-26, and
later to AKL-26A; serial # 2-29 and # 2-30
later modified to AKL-26A; serial # 2-32
believed to be NC-101M; several serial numbers left unverified.

A.T.C. #204
(8-23-29)
AEROMARINE-KLEMM, AKL-26A

Fig. 12. Aeromarine-Klemm model AKL-26A with 60 H.P. Le Blond engine.

Shown here in various views, the Aero-marine-Klemm AKL-26A was a companion model to the LeBlond powered AKL-26 de-scribed here just previous, and was basically typical in most all respects. To increase the operating range of the AKL, an extra fuel tank of 14 gal. capacity was mounted in the right hand side of the wing stub; this prac-tically doubled the cruising radius. To carry this extra fuel and still carry the same basic payload, the AKL-26A version was operated at a higher gross weight; because of this, the landing speed was slightly higher and the rate of climb-out was proportionately less. A good bit friskier than the 40 h.p. model AKL-25A, this new series of the Aeromarine-Klemm monoplane was noted as a pleasure craft "with a capital P"; for sheer going-nowhere flying fun, this was one of the best and a good number of these delightful craft were still flying some 10 years later.

The Aeromarine-Klemm model AKL-26A as powered with the 5 cyl. LeBlond 5-D of 65 h.p., received its type certificate number on 8-23-29 for both a landplane and seaplane version. The seaplane version was fitted with either Edo model H or Aeromarine-Kantner twin float gear; as pictured here, the AKL-26A was also tested with an unusual set of amphibious pontoons. Existing examples of the AKL-25A and AKL-26 were also eligible for certification under ATC # 204 as model AKL-26A when modified to conform. A year or so later, both the AKL-26 and AKL-26A were allowed the installation of the improved LeBlond 5-DE of 70 h.p., which increased the all-around performance to a slight degree. In all, some 30 or more examples of the AKL-26A version were manufactured by the Aero-marine-Klemm Corp. at their well-known plant in Keyport, New Jersey.

Listed below are specifications and per-formance data for the Aeromarine-Klemm model AKL-26A as powered with the 65 h.p. LeBlond 5-D engine; length overall 23'6"; hite overall 7'0"; wing span 40'2"; wing chord at root 79"; total wing area 194.5 sq.ft.; airfoil Goettingen 387 modified; wt. empty (land-plane) 1025; useful load 565; payload with

29 gal. fuel was 210 lbs.; gross wt. 1590 lbs.; max. speed 93; cruising speed 80; landing speed 40; climb 400 ft. first min. at sea level; ceiling 9000 ft.; gas cap. 29 gal.; oil cap. 2 gal.; cruising range at 4.5 gal. per hour was 480 miles; price at the factory was $3550. The following figures are for twin-float seaplane; wt. empty 1100 lbs.; useful load 500; payload with 29 gal. fuel was 145 lbs.; gross wt. 1600 lbs.; max. speed 85; cruising speed 75; landing speed 40; climb 375 ft. first min. at sea level; ceiling 8500 ft.; cruising range 450 miles; price at the factory was $4500. and up. As is most always the case, the earliest versions of the AKL-26A were a good bit lighter; wt. empty 954; useful load 576; payload with 29 gal. fuel was 221 lbs.; gross wt. 1530 lbs.; for early seaplane version as follows; wt. empty 1090; useful load 500; payload 145 lbs.; gross wt. 1590 lbs.; all other figures were typical, and performance was proportionately better.

The angular fuselage framework was built up of spruce longeron members, plywood bulkheads, and formers; the completed framework was covered in plywood veneer to form a semi-monococque shell. The engine mount was a welded steel tube, detachable framework; a baggage compartment of 6 cu. ft. capacity, with allowance for 36 lbs., was mounted in the turtle-back section of the fuselage, behind the rear cockpit. The cantilever wing framework, in three sections, was built up of spruce and plywood box-type spar beams and plywood wing ribs; the completed framework was covered in plywood veneer. Outer wing panels were of tapering chord and thickness, attached to a center-section panel of constant chord and thickness. A gravity-feed fuel tank of 15 gal. capacity was mounted high in the fuselage just ahead of the front cockpit; an extra 14 gal. fuel tank was mounted in the right hand side of the wing stub. The split-axle landing gear of 85 inch tread, was an inverted tripod affair using oleo-spring shock absorbing struts; wheel brakes were available and 20x9 low pressure "air-wheels" were optional. The fixed surfaces of the tail-group were built up of spruce spars and ribs, covered in plywood veneer; the movable surfaces were also built up of spruce spars and ribs, covered in fabric. The vertical fin and the horizontal stabilizer were ground adjustable only; dual stick controls were provided. Both the AKL-26 and the AKL-26A were later approved with a tail bracing strut attaching to the vertical fin and horizontal stabilizer; this strut was of streamlined steel tubing. The next development in the Aeromarine-Klemm monoplane was the AKL-26B, powered with 85 h.p. LeBlond

Fig. 13. Gentle nature of AKL-26A ideal for pilot training.

*Fig. 14. Rugged structure offered hours
of care-free flying.*

engine, which will be discussed in the chapter
for ATC # 334.

Listed below are AKL-26A entries that
were gleaned from various records; this is
not a complete listing but it does show the
bulk of this model that were built:

NC-122H; AKL-26A (# 2—16) LeBlond 60
NC-123H;   „   (# 2—17)   „
NC-161H;   „   (# 2—19)   „
NC-52K;   „   (# 2—29)   „
NC-100M;   „   (# 2—30)   „
NC-102M;   „   (# 2—33)   „
NC-161M;   „   (# 2—36)   „

NC-162M;   „   (# 2—37)   „
NC-195M: „   (# 2—40)   „
NC-196M;   „   (# 2—41)   „
NC-197M;   „   (# 2—43)   „
NC-39K;   „   (# 2—44)   „
NC-199M;   „   (# 2—47)   „
NC-312N;   „   (# 2—48)   „
NC-313N;   „   (# 2—49)   „
NC-163M;   „   (# 2—50)   „
NC-54K;   „   (# 2—52)   „
NC-159M;   „   (# 2—55)   „
NC-318N;   „   (# 2—57)   „
NC-319N;   „   (# 2—58)   „
NC-320N;   „   (# 2—59)   „
NC-321N;   „   (# 2—60)   „
NC-316N;   „   (# 2—61)   „
NC-388N;   „   (# 2—63)   „
NC-391N;   „   (# 3—66)   „
NC-861W;   „   (# 3—70)   „
NC-865W;   „   (# 4—76)   „

Serial # 2-50 (NC-163M) first as AKL-26A
with LeBlond 60 later modified with installation of 80 h.p. "Genet" engine; several other
serial numbers unverified, which could have
been AKL-26A models; several Salmson-
powered AKL-25A were later modified to
AKL-26A or AKL-16A specifications.

Fig. 15. Aeromarine-Klemm in test with amphibious float gear.

A.T.C. #205
(8-17-29)
TRAVEL AIR, K-4000

Fig. 16. K-4000 another example of popular Travel Air biplane.

The Kinner powered model K-4000 was one of the rare and seldom-seen versions in the Travel Air biplane series; a delightful engine-airplane combination that would have surely found particular favor amongst the private-owner flyer or the small operator, had it been released for sale to the flying public about a year or so sooner. With the huge supply of Curtiss OX-5 engines dwindling away at a rapid rate, many of the manufacturers were now grooming models with low-powered aircooled engines but the choice of powerplants in this range was still quite limited; the cost of these new engines was very high per horsepower in comparison to the OX-5, which was built cheaply in great number and then bought up from war-surplus stocks. All these factors had a direct bearing on the selling price of the new model aircraft which averaged out about $5000. at the factory; an OX-5 powered airplane of the same type could be bought for about half the price. Several good low-powered foreign engines were available in this power range, but due to a number of circumstances they proved to be more of a bother than a boon. Travel

Air had introduced the friendly and mild-mannered model W-4000 earlier in the year, which was a fine craft and certainly selling very well, but the introduction of the Kinner powered K-4000 came a little too late and at a rather bad time; consequently only a small number were built and it remained a rather rare type.

The model K-4000 as shown here in very good likeness, was a typical 3 place open cockpit Travel Air biplane of a familiar configuration that was coming off the production line at this particular time. Powered with the 5 cyl. Kinner K5 engine of 100 h.p., its performance was quite adequate and its economy of maintenance and operation were its cardinal features. Due to their scarcity, very little lore is known about this particular model, but it has been reported on occasion that this was an airplane of compatible nature and a happy combination that readily adapted itself to the average chores at hand. Its rugged structural dependability and its marked operating reliability were typical built-in attributes that made the Travel Air biplane such a great

Fig. 17. Famous Travel Air "Mystery Ship" was revolutionary design that paved way for new concepts in aeronautical engineering.

favorite the country over. The Kinner Airplane & Engine Co. placed considerable faith in the future of this new Travel Air, and used a K-4000 extensively as a demonstrator to promote sales of their K5 engines; most other examples of this model were owned by private owner flyers. The approved type certificate for the model K-4000, as powered with the 100 h.p. Kinner K5 engine, was issued 8-17-29 and some 6 or more examples of this model were manufactured by the Travel Air Co. at Wichita, Kan.

Listed below are specifications and performance data for the model K-4000 as powered with the Kinner K5 engine; length overall 24'8"; hite overall 8'11"; wing span upper 33'0"; wing span lower 28'10"; wing chord upper 66"; wing chord lower 56"; total wing area 289 sq.ft.; airfoil T.A. # 1; wt. empty 1340; useful load 940; payload with 42 gal. fuel was 478 lbs.; gross wt. 2280 lbs.; max. speed 100+; cruising speed 88; landing speed 43; climb 580 ft. first min. at sea level; ceiling 10,000 ft.; gas cap. 42 gal.; oil cap. 5 gal.; cruising range at 7 gal. per hour was 425 miles; price at the factory field averaged at $5000. with extra equipment.

The fuselage framework was built up of welded chrome-moly steel tubing, and Travel Air still held to the fashion of bracing the rear portion of the framework with steel tie rods; the framework was heavily faired to shape with wood fairing strips and then fabric covered. The wing framework was built up of laminated spruce spar beams that were routed out for lightness, with spruce and plywood built-up wing ribs; the completed panels were fabric covered. The cockpits were deep and spacious with provisions for baggage in a quickly accessible compartment behind the rear cockpit, with a capacity of about 9 cu. ft.; the fuel tank was mounted high in the fuselage, just ahead of the front cockpit, with a direct-reading fuel gauge projecting up thru the cowling. Interplane struts were of streamlined chrome-moly steel tubing and interplane bracing was of streamlined steel wire; ailerons in the upper wing only, were actuated by a push-pull strut. The landing gear was of the typical split-axle type and used two spools of shock-cord to snub the bumps; wheels were 28x4 and brakes were available. The fabric covered tail-group was built up of welded chrome-moly steel tubing and sheet former ribs; the fin was ground adjustable and the horizontal stabilizer was adjustable in flight. A metal propeller, navigation lights, and an air-operated Heywood engine starter were available as optional equipment. The next development in the Travel Air biplane series was the model 4-D, which is discussed in the chapter for ATC # 254 in this volume.

Listed below are Travel Air model K-4000 entries that were gleaned from registration records:

NC-8841; K-4000 (# 1005) Kinner K5
NC-9962;   „   (# 1161)   „
NC-9963;   „   (# 1162)   „
NC-9987;   „   (# 1163)   „
NC-464N;   „   (# 1359)   „
NC-453N;   „   (# 1373)   „

NC-8896 was serial # 952, first built as 2000 with OX-5, later modified to K-4000 with installation of Kinner K5 engine, operated in Hawaiian Islands.

BOEING TRI-MOTOR, 80-A

Fig. 18. Boeing model 80-A high above clouds in flight across country.

The Boeing model 80-A air-liner was designed primarily for large capacity transcontinental passenger service or combined passenger and mail-express operation. There were accommodations for 18 passengers in the 80-A version and seating for 12 passengers, plus mail and cargo, in the 80-A1 version. Eleven of these huge tri-engined transports were in use by Boeing Air Transport which inaugurated daily round-trip service on a 20 hour schedule between Chicago and San Francisco in May of 1930; a rugged route which posed extremes of temperature, weather, terrain, and altitudes. The first night-flights on regular schedule were also inaugurated with the 80-A transport by this line, on a route between Salt Lake City and Oakland, Calif.; a trip of 634 miles that took only one-quarter the time of regular train service. Designed for combined operations, the passenger chairs could be removed in about 15 minutes and the cabin interior converted into a flying post-office, where mail could be sorted and routed while in flight. After first using "male couriers" on Boeing 80-A flights, it was decided to use women as "stewardess"; 8 specially trained nurses were of the first group to be used on daily flights

from the Golden Gate to the shores of Lake Michigan. This service was later extended to New York City on a 27 hour schedule.

With seating for 18 passengers and a crew of 2, the huge 80-A transport was the outstanding big-airplane attraction at the Cleveland Air Show of 1929; it was too large to be housed in the show building so it was on display outdoors, where it attracted a constant stream of interested on-lookers. Basically similar to the model 80 transport of 1928, the new 80-A was powered with 3 Pratt & Whitney "Hornet" B engines of 525 - 575 h.p. each, to allow for greater capacity and better performance; the new improved model contained many advances in its design and construction that were developed by 7 million miles of flying experience by the Boeing Air Transport System. The main cabin, lined with 3 rows of seats, was tastefully finished in polished mahogany plywood with a thick core of balsa wood for insulation. Soundproofing of the cabin walls allowed conversation in near-normal tones, and heating and ventilation kept cabin temperatures at a fairly comfortable level; featured also was a cloak-room, overhead racks for hats and small

Fig. 19. Boeing 80-A coming into Chicago on flight from San Francisco.

baggage, and a small lavatory with hot and cold running water. Baggage allowance was up to 50 lbs. per passenger, which was stored in a 39 cubic foot compartment below the floor of the pilot's cabin; the combination mail and passenger version carried 12 passengers and over 1400 lbs. of mail and air-express cargo. With a loaded gross weight of 8 and three-quarter tons, this was a huge airplane for its day and one of the finest transports of this early era.

Stemming from the earlier Wasp-powered model 80, which was already in a test program by August of 1928, the improved model 80-A was introduced in September of 1929. Continuous development saw this airplane blossom out into several versions; the prototype version had 3 cowled-in engines; another version had 1 bare engine and 2 cowled-in engines; some had 3 bare engines without any low-drag cowlings; and all were later converted to model 80-A1 specs which had provisions for more fuel, carried 12 passengers and cargo-freight, and had rudder area added in the form of 2 sub-rudders. The type certificate number for the model 80-A and 80-A1 was first issued 8-20-29; some 12

examples of this model were manufactured by the Boeing Airplane Co. at Seattle, Wash., which by now employed a growing force of over 800 people. Philip G. Johnson was the president; C. L. Egtvedt was V.P. and general manager; and C. N. Montieth was chief of engineering. The Boeing company was now a division of the United Aircraft & Transport Corp.

Listed below are specifications and performance data of the Boeing model 80-A as powered with 3 "Hornet" engines of 525 h.p. each; length overall 56'6"; hite overall 15'3"; wing span upper 80'0"; wing span lower 64'10"; wing chord upper 120"; wing chord lower 88"; wing area upper 800 sq.ft.; wing area lower 450 sq.ft.; total wing area 1250 sq. ft.; airfoil Boeing 106; wt. empty 10,582; useful load 6918; payload with 388 gal. fuel was 3962 lbs.; gross wt. 17,500 lbs.; max. speed 138.7; cruising speed 118; landing speed 61; climb 900 ft. first min. at sea level; climb in 10 min. was 5500 ft.; ceiling 14,000 ft.; gas cap. 388 gal.; oil cap. 36 gal.; cruising range at 90 gal. per hour was 460 miles; the following figures are for the 12 passenger 80-A1 which also carried mail and cargo; wt. empty

10,735; useful load 6765; payload with 392 gal. fuel was 12 passengers and 1145 lbs. of cargo; gross wt. 17,500 lbs.; performance figures remained more or less the same; fuel cap. for both versions was 400 gal. max.; price at the factory was $75,000.

The fuselage framework was built up of welded chrome-moly steel tubing for the entire nose-section up to the rear wing spar attach point, and other highly stressed areas; the balance of the fuselage was built up of square dural tubes that were bolted together to form the trussing, and the completed framework was fabric covered. The fuselage trusses were designed for unobstructed window openings and the cabin interior was lined with polished mahogany; the cabin walls were lined with a thick core of balsa-wood for sound-proofing and insulation. Provision was made for heating and ventilation to each seat, and the sound-proofed walls kept noise at a fairly low level; reading lights were provided by every seat and all windows were of shatter-proof glass. All passenger chairs were attached with quick-lock fittings to allow easy removal or change in seating arrangement; seats in the forward compartment could by removed to allow space for up to 1400 lbs. of cargo and mail. The wing framework was built up of two spar beams that were deep trusses built up of square dural tubing that was bolted together; the wing ribs were riveted dural tube and channel section trusses, and the completed framework was fabric covered except for the metal walk-way on the root ends of the lower wings. Engine nacelle structure and the inboard interplane struts were of chrome-moly steel tubing; outer wing struts were of streamlined dural tubing and the interplane bracing was of streamlined steel tie-rods. All fuel was carried in the center section panel of the upper wing; there were two tanks of 165 gal. capacity each and a 70 gal. reserve tank.

Fig. 20. Boeing model 80 of 1928 pioneered flights paving way for transcontinental system.

Oil supply was carried in 3 tanks, one behind each engine; a cockpit-operated fire extinguisher was placed in each engine nacelle. Fixed surfaces of the tail-group were of metal construction and covered in corrugated dural skin; the rudder and elevators were built up of welded chrome-moly steel tubing and covered in fabric. There were several variations in rudder configuration on the 80-A at different times, but the final shape was a triple-rudder arrangement which contained a semi-servo booster to overcome heavy pressures, especially in the failure of either outboard engine; the horizontal stabilizer was adjustable in flight to affect trim. The treadle type landing gear of 18′2″ tread had oleo shock absorbing struts and Bendix wheels and brakes; wheels were 54x12 and a full-castor tail wheel was 15x3. There was an electric inertia-type starter for each engine, and the framework of the airplane was bonded and shielded for radio. The next Boeing development was the model 203-A trainer which will be discussed in the chapter for ATC # 211 in this volume.

Listed below are Boeing 80-A entries that were gleaned from Boeing Airplane Company records:

NC-793K;	Model 80-A	(# 1081)	3 Hornet
NC-224M;	„ „	(# 1082)	„
NC-225M;	„ „	(# 1083)	„
NC-226M;	„ „	(# 1084)	„
NC-227M;	„ „	(# 1085)	„
NC-228M;	„ „	(# 1086)	„
NC-229M;	„ „	(# 1087)	„
NC-230M;	„ „	(# 1088)	„
NC-231M;	„ „	(# 1089)	„
NC-232M;	„ „	(# 1090)	„
NC-233M;	„ „	(# 1091)	„
NC-234M;	„ „	(# 1092)	„

Serial # 1091 later converted to the "Model 226" (group 2-248) which was a deluxe executive type craft built for Standard Oil Co.; serial # 1092 first built as 80-B with open cockpits for pilot, later rebuilt to 80-A1 specs; all of the craft listed above with the exception of serial # 1083 and # 1091 were later converted to 80-A1 specs. The earlier Boeing 80, of which 4 were built, was on Group 2-4; the model 80-C was a proposed freight-carrying version of the 80-A but none were ever built.

Fig. 21. Boeing 80-A coming in from the east; shown here over Sacramento Valley.

Fig. 22. Model 80-A1 carried 12 passengers and large load of mail cargo.

Fig. 23. Cessna DC-6 with Curtiss Challenger engine was first of new series.

With the model AW production line humming along at a satisfying rate, Clyde V. Cessna (still looked upon and revered as the "old master") turned to the development of several larger craft he had in mind; Cessna's first move was the development of a large 6 place cabin monoplane. On November 1 of 1928, a shiny new airplane was rolled out of the final assembly building; it was an impressive craft that was patterned after traditional Cessna cantilever-winged design and had seating capacity for 6. This was the model CW-6 which was powered with a 220 h.p. Wright "Whirlwind" J5 engine; pure white and dolled up pretty with bright red trim, this craft was proudly exhibited at the 1929 Auto Show in Wichita.

Enticed by representatives from the huge country-wide Curtiss Flying Service organization to expand production facilities and offer new models, a new enlarged plant (some 3 times larger than the old plant) was erected on 80 acres nearby and new models were designed to fit into the rosy picture of business yet to come. Riding high on a wave of buying that spiralled ever-upward, the Curtiss sales organization offered to handle Cessna's complete out-put, promising to sell 50 airplanes per month. It was hard to reject this offer of on-coming prosperity, but Clyde Cessna surely had tried.

To meet these new contracts and to supplement the "Cessna AW" line with a larger 4 place model, the DC-6 was designed as a baby version of the earlier CW-6. More or less obligated to be powered with a 6 cyl. Curtiss "Challenger" engine of 170 h.p., this new craft was rolled out for flight in February of 1929. After flight tests and thorough performance evaluation, it was decided that this new model was basically sound and a definite step in the right direction but it did need an increase in horsepower to offer the performance that the flying public was coming to expect. Almost immediately this design was modified and groomed for the 7 cyl. and 9 cyl. Wright J6 series engines, which were combinations expected to top anything aloft in this class of airplane; these 2 new models were the DC-6A and the DC-6B. Because of the quick decision to increase horsepower, the Cessna DC-6 as powered with the "Challenger" engine was built in only one example to remain as the forerunner of a new line of airplanes by Cessna. Clyde V. Cessna was still the president of the Cessna Aircraft Co. and

Fig. 24. Prototype of DC-6 was like scaled-down version of CW-6.

in charge of design and engineering, but he was now grooming his son Eldon, as the chief engineer and sales manager; the colorful "Chief" Bowhan was the chief pilot in charge of test and development.

Listed below are specifications and performance data for the Cessna model DC-6 as

powered with the 170 h.p. Curtiss "Challenger" engine; length overall 27'11"; hite overall 7'8"; wing span 40'8"; wing chord 78" mean; total wing area 265 sq.ft.; airfoil "Cessna" (modified M-12); wt. empty 1767; useful load 1221; payload with 62 gal. fuel was 633 lbs.; gross wt. 2988 lbs.; max. speed 130; cruising speed 105; landing speed 48; climb 780 ft.

Fig. 25. CW-6 on display at auto show in Wichita, Kan.

Fig. 26. DC-6 was enlarged version of popular Cessna AW.

first min. at sea level; ceiling 16,000 ft.; gas cap. 62 gal.; oil cap. 5 gal.; cruising range at 10 gal. per hour was 575 miles; price at the factory field was $9250.

The fuselage framework was built up of welded chrome-moly steel tubing, faired to shape with plywood formers and wooden fairing strips; the completed framework was fabric covered. The cantilever wing framework was built up of spruce box-type spar beams and spruce and plywood truss-type wing ribs; the leading edge was covered to the front spar to preserve the airfoil form and the completed framework was covered with fabric. Narrow-chord ailerons were of the "Friese" balanced-hinge type, and the two fuel tanks were mounted in the wing flanking the fuselage. The cabin interior had ample room for four and an 18 cubic foot baggage compartment was located behind the rear seat. The long-leg landing gear was first attached to the upper longeron of the fuselage but upper attach points were later moved out to the front spar

of the wing, as shown in one of the illustrations; wheel tread was 102 inches and 30x5 wheels were fitted with Bendix brakes. The fabric covered tail-group was built up of welded chrome-moly steel tubing; the fin was ground adjustable and the horizontal stabilizer was adjustable in flight. The next development in the Cessna DC-6 series were the two more powerful versions that are discussed in the chapters for ATC # 243 and # 244 in this volume.

The only example of the "Challenger" powered Cessna model DC-6 is listed below: Registration number NC-9867; serial number unknown, but it more than likely was # 157. There may have been two airplanes of this model but more than likely, as Cessna Aircraft records suggest, the prototype was registered with identification number of 8142 in its original version as shown, and then registered as NC-9867 when modified to its final configuration.

A.T.C. #208
(8-20-29)
FAIRCHILD, KR-34A

Fig. 27. Rare KR-34A was typical of Kreider-Reisner 34-series with 6 cyl. Curtiss Challenger engine.

The Kreider-Reisner KR-34 ("Challenger" C-4) was a very successful series that mounted several different power-plants to achieve the various models; the C-4A or KR-34A that is under discussion here, was a rather rare version in this familiar line. The model KR-34A was also a 3 place open cockpit biplane of the all-purpose type and was basically similar to the KR-34B or KR-34C, except for its engine installation which in this case was the 6 cyl. Curtiss "Challenger" twin-row radial air-cooled engine of 170-185 h.p. The comparative scarcity of the model KR-34A made it unfamiliar to many and therefore not much lore remains about its performance or general behavior; other versions of the KR-34 biplane were certainly top-notch airplanes, so we can safely assume that this model also shared the many fine qualities that made the KR-34 line a country-wide favorite. The type certificate number for the model KR-34A, as powered with the 170 h.p. Curtiss "Challenger" engine, was issued 8-20-29. A former Kreider-Reisner "Challenger" design (the name having no connection with the "Challenger" engine) that was now built under the name of Fairchild, the Fairchild

KR-34A was built in some 2 or more examples by the Kreider-Reisner Aircraft Co., Inc. of Hagerstown, Md., a subsidiary of the Fairchild Airplane Mfg. Corp. In April of 1929, the Kreider-Reisner firm was absorbed by Fairchild and the KR biplanes were added to their line of monoplanes; instead of being called Kreider-Reisner "Challenger" biplanes, they were now called Fairchild "KR" biplanes. John S. Squires was the president of the KR division, and veteran Louis E. Reisner was the V.P.; Amos Kreider, one of the original founders, had retired to other interests.

Listed below are specifications and performance data for the Fairchild model KR-34A as powered with the 170 h.p. Curtiss "Challenger" engine; length overall 23'2"; hite overall 9'3"; wing span upper 30'1"; wing span lower 28'9"; wing chord both 63"; wing area upper 154 sq.ft.; wing area lower 131 sq.ft.; total wing area 285 sq.ft.; airfoil "Aeromarine" 2-A modified; wt. empty 1462; useful load 901; payload with 53 gal. fuel was 373 lbs.; gross wt. 2363 lbs.; max. speed 122; cruising speed 104; landing speed 45; climb 886 ft. first min. at sea level; climb in 10 min. was

6,000 ft.; ceiling 14,500 ft.; performance with 185 h.p. "Challenger" engine was proportionately better; gas cap. 53 gal.; oil cap. 5 gal.; cruising range at 10.5 gal. per hour was 460 miles; price at the factory averaged $6575.-$7000., depending on equipment requested.

The fuselage framework was built up of welded chrome-moly steel tubing, faired to shape with wooden fairing strips and fabric covered; a small baggage compartment of 35 lb. capacity was placed in the fuselage behind the rear cockpit. A door was provided for entrance to the front cockpit and a convenient step was provided for easy entrance to the rear cockpit. The wing framework was built up of solid spruce spar beams that were routed out for lightness, and wing ribs were built-up trusses of spruce and plywood; the completed framework was fabric covered. The gravity-feed fuel tank was mounted high in the fuselage just ahead of the front cockpit; an oil tank was mounted in the engine section. The split-axle landing gear of 72 inch tread employed rubber discs in compression as shock absorbers, but oleo-spring shock absorbing struts were later available; wheels were Bendix 28x4 or 30x5 and brakes were optional equipment. The fabric covered tail-group was built up of welded chrome-moly steel tubing; the fin was ground adjustable and the horizontal stabilizer was adjustable in flight.

Other standard equipment included metal propeller, dual stick controls, cockpit covers, booster magneto, tie-down ropes, log books, tool kit, first aid kit, fire extinguisher, wiring for navigation lights, engine cover, maintenance booklet, and an above-average set of engine and flight instruments. The next development in the Kreider-Reisner biplane was the C-6B sport-trainer (KR-21A), which is discussed in the chapter for ATC # 215 in this volume.

Listed below are Fairchild KR-34A entries that were gleaned from registration records; as far as can be determined, this was the total production of this model:

NC-567K ; KR-34A (# 356) CurtissChallenger
NC-11607;   „   (# 364)   „   „

In checking serial numbers from # 356 to # 364, the following were KR-31, # 357-359-361-362; the only possibles for other KR-34A examples would be serial # 358-360-363 but none were registered.

A.T.C. #209
(8-22-29)
COMMAND-AIRE, 3C3-BT

Fig. 28. Command-Aire trainer model 3C3-BT powered with 113 H.P. Yankee-Siemens (Siemens-Halske) engine.

The next progressive development in the popular Command-Aire trainer series was the model 3C3-BT, which was typical to previous models (3C3-T and 3C3-AT) except for the powerplant installation which in this case was the German-built Yankee-Siemens (Siemens-Halske) SH-14 engine of 105-113 h.p. The rare and seldom seen 3C3-BT trainer was also a two-place open cockpit biplane seating two in tandem; a compatible engine-airplane combination that shared all the appealing qualities and flight characteristics that had made the Command-Aire biplane a country-wide favorite for many years. Turning to markets abroad for additional sales, a Command-Aire demonstration team toured countries of South America in search of likely prospects; after a successful demonstration, Command-Aire, Inc. received an order from the Chilean government for 36 of the 3C3-BT traing planes powered with Siemens-Halske engines. The Siemens-Halske was chosen as a powerplant for the 3C3-BT because it was more apt to appeal to customers in the foreign market. Such an order would have indeed been quite a "plum" for an American aircraft manufacturer, but no confirming records have been available to show what was

delivered, or if the order ever went through at all.

Command-Aire airplanes in all models had been sold and distributed solely by the widespread Curtiss Flying Service organization, but towards the latter half of 1929, the company decided to develop its own sales organization under the direction of Maj. J. Carroll Cone. The model 3C3-BT trainer as powered with the 7 cyl. Yankee-Siemens engine of 105-113 h.p., received its type certificate number on 8-22-29 and some 3 or more examples of this model were manufactured by Command-Aire, Inc. at Little Rock, Arkansas. R. B. Snowden Jr., a true sportsman-pilot, was the president; Albert Voellmecke, talented European aircraft designer, was the chief engineer; J. Carroll Cone, former Air Corps pilot, was sales manager; and Wright "Ike" Vermilya, the fearless one, was the chief test pilot.

Listed below are specifications and performance data for the Command-Aire trainer model 3C3-BT as powered with the 105-113 h.p. Yankee-Siemens engine; length overall 24'6"; hite overall 8'6"; wing span upper & lower 31'4"; wing chord both 60"; wing area upper 156 sq.ft.; wing area lower 146 sq.ft.; total wing area 302 sq.ft.; airfoil "Aero-marine" 2-A; wt. empty 1340; useful load 682; payload with 40 gal. fuel was 232 lbs.; gross wt. 2022 lbs.; max. speed 110; cruising speed 92; landing speed 38; climb 720 ft. first min. at sea level; ceiling 13,000 ft.; gas cap. 40 gal.; oil cap. 5 gal.; cruising range at 6 gal. per hour was 550 miles; price at the factory averaged $5600., then lowered to $3350. to create more sales in mid-1930.

Typical of other Command-Aire aircraft the fuselage framework of the 3C3-BT was built up of welded chrome-moly steel tubing, faired to shape with fairing strips and fabric covered. The wing framework was built up of solid spruce spar beams and wing ribs were built up of spruce and plywood; the leading edge was covered with plywood and the completed framework was fabric covered. The interior was kept rather bare of frills and fancy appointments for the sake of maintenance and ease of inspection; instruments provided were kept to a minimum, dual stick controls were provided, and seats were shaped with deep wells for parachute packs. The shock-cord sprung landing gear was of 84 inch tread; wheels were 28x4 or 30x5. Wheel brakes, metal propeller, navigation lights, and inertia-type engine starter were available. All other details of construction and arrangement

Fig. 29. 3C3-BT was typical of Command-Aire trainer series, sharing inherent good nature and performance.

typical of all models of the Command-Aire trainer series are covered more fully in U.S. CIVIL AIRCRAFT, Vol. 2 in the discussion for ATC # 151 which covers the model 3C3-AT. The next development in the Command-Aire biplane was the Axelson-powered model 5C3-B which is discussed in the chapter for ATC # 214 in this volume.

Listed below are Command-Aire model 3C3-BT entries that were gleaned from registration records:

NC-906E; 3C3-BT (# W-70) Yankee-Siemens.
NC-952E;   „   (# W-97)   „   „
NC-958E;   „   (# W-104)   „   „
Serial # W-104 first as 3 place model 3C3-B, later modified to model 3C3-BT.

Fig. 30. Command-Aire "Little Rocket" was winner of Cirrus Derby.

Fig. 31. Aristocrat 102-E with 165 H.P. Wright J6 engine.

A general improvement had been planned for the tidy-looking "Aristocrat" cabin monoplane, and the basis of this general refinement revolved mostly around the increase in horsepower. Formally announced as the model 102-E, this new craft was basically a progressive development and quite similar to the Warner-powered 102-A, except for the powerplant installation which in this case was the 5 cyl. Wright J6 "Whirlwind Five" engine of 165 h.p.; performance and general utility was substantially improved with the addition in horsepower. Several of this new model were used in charter air-taxi work and a few were used by business-houses in sales development; "Air Associates", an airplane parts supply house, had a 102-E fitted especially as a "flying show-case" to demonstrate their wares in all parts of the country. First brought into some prominence when 8 of the "Aristocrat" 102-A were demonstrated by the General Tire & Rubber Co. on a 50,000 mile nation-wide tour that also swung through Mexico and Cuba, these efficient monoplanes helped to prove the reliability of airplanes in cross-country air travel to all points of the compass, under all manner of conditions. At the same time the tour proving conclusively that the basic "Aristocrat" design had plenty of muscle and utility; the performance of this craft in any model was excellent, and flight characteristics were pleasant and predictable.

The General "Aristocrat" model 102-E as shown, was a 3 place high-wing cabin monoplane basically similar to the model 102-A, except for the powerplant installation which in this case was the 5 cyl. Wright J6 (R-540) engine of 165 h.p. Neat and quite trim, the model 102-E seated a pilot and two passengers in ample room and good comfort; general utility was enhanced and the performance had now been substantially improved with the addition of more horsepower. Operating costs of this new model were now slightly higher, but the increase in general usefulness was well worth the added price. The "Aristocrat" model 102-E as powered with the 165 h.p. Wright J6 engine, received its type certificate number on 8-23-29 and some 6 or more examples of this model were manufactured by the General Airplanes Corp. at Buffalo, New York. C. S. Rieman was president and general manager; L. A. Listug was V.P. and treasurer; A. Francis Arcier was V.P. in charge of engineering and development; and George A. Townsend was the secretary and director of sales. The engineering staff was headed by A. Francis Arcier whose experience was noted both here and abroad; Arcier later presided as chief engineer for Waco Aircraft after manufacture of General airplanes was discontinued in the early thirties.

Listed below are specifications and performance data for the General "Aristocrat" model 102-E as powered with the 165 h.p. Wright J6 engine; length overall 26'6"; hite overall 8'0"; wing span 36'8"; wing chord 72"; total wing area 198 sq.ft.; airfoil GAC-500; wt. empty 1524; useful load 776; payload with 40 gal. fuel was 343 lbs.; gross wt. 2300 lbs.; max. speed 128; cruising speed 110; landing speed 48; climb 750 ft. first min. at sea level; climb in 10 min. was 6000 ft.; ser.

ceiling 16,700 ft.; gas cap. 40 gal.; oil cap. 3.5 gal.; cruising range at 9.5 gal. per hour was 480 miles; price at the factory was originally quoted at $7500. and then lowered to $5250. in 1930.

The fuselage framework was built up of welded chrome-moly steel tubing in truss form, faired to shape with formers and fairing strips and then fabric covered. The pilot sat up front on a bucket-type seat, and the two passengers sat on a bench-type seat in back; windows were of shatter-proof glass and could be rolled up and down for cabin ventilation. There was a wide door on each side for cabin entry, and a baggage compartment of 7 cubic foot capacity behind the rear seat; no baggage was allowed with 3 passengers when loaded to total of 508 lbs. The rear seat was quickly removable for the hauling of cargo. The wing framework of semi-cantilever design, was built up of laminated spruce spar beams, with spruce and plywood truss-type wing ribs; the leading edge was covered to the front spar with duralumin sheet to preserve the airfoil form and the completed frame-

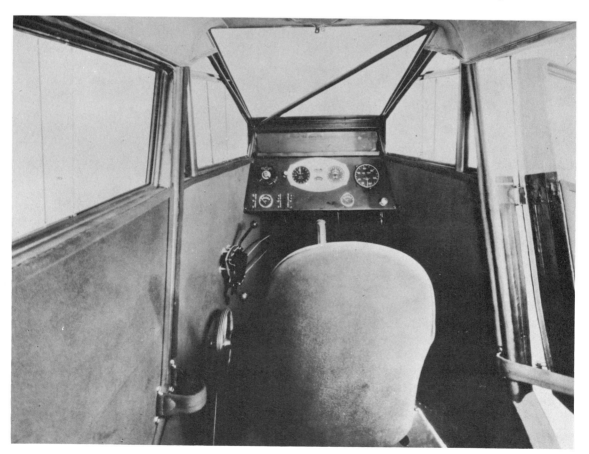

Fig. 32. Aristocrat interior showing pilot station.

Fig. 33. Aristocrat with "Whirlwind Five" offered better performance.

work was fabric covered. The ailerons were built up of chrome-moly steel tube spar beams and dural sheet ribs, covered with fabric; for more efficient control, the ailerons were of the aerodynamically balanced offset-hinge type. Two gravity-feed fuel tanks were mounted in the root ends of each wing half, one either side of the fuselage; wing braces were large diameter steel tube struts of streamlined section. The fabric covered tail-group was built up of welded chrome-moly steel tubing, with both rudder and elevators fitted with aerodynamically balanced "horns"; the vertical fin was ground adjustable and the horizontal stabilizer was adjustable in flight. The split-axle landing gear of 84 inch tread was built up of two heat-treated aluminum alloy cantilever legs that were faired to a streamlined shape by sheet metal cuffs; the shock absorbers were rubber "donut rings" in compression. Main wheels were 30x5 Bendix; individual wheel brakes and a swiveling tail-wheel provided excellent ground maneuvering. The engine was well muffled by a large volume exhaust collector-ring that permitted normal conversation in the cabin at all times. Flight controls were of the joy-stick type and a dual set was available; Bendix wheels and brakes, a metal propeller, a hand-crank in-

ertia-type engine starter, and navigation lights were standard equipment. The "Aristocrat" model 102-E was called the 102-B when first proposed with the early "Wright Five" of 150 h.p.; the models 102-C and 102-D were rumored to have been proposed models with Axelson and Comet engines.

The next development in the "General Monoplanes" series was the parasol-winged "Cadet" trainer which is discussed in the chapter for ATC # 229 in this volume.

Listed below are "Aristocrat" model 102-E entries that were gleaned from registration records:

NC-281H; 102-E (# 23) J6-5-165
NC-282H;   „   (# 24)   „
NC-280H;   „   (# 25)   „
NC-284H;   „   (# 26)   „
NC-715Y;   „   (# 30)   „
NC-492K;   „   (# 41)   „
NC-493K;   „   (# 42)   „

Serial # 30 later modified as 102-F with 165 h.p. Continental A-70 engine; identity of serial number 33 and 34 unknown, probably as model 102-E; serial # 43 was built in 1931 and listed both as 102-B and 102-E, registered X-11311.

Fig. 34. Boeing model 203-A trainer with 165 H.P. Wright J6 engine.

The Boeing model 203 series biplane was unusual only in the fact that it was especially designed as a pilot-training airplane for use at the Boeing School of Aeronautics, based at Oakland, Calif., and not for sale on the civilian market. Born and reared among such other famous Boeing airplanes as the "Model 95" and the "Model 100", some of the shape and characteristics were bound to show up in its make-up and that is why it plainly reflected some traits of each. Conforming to accepted practice for a training airplane, the "203" was an open cockpit biplane of rather good nature and very rugged character. Designed primarily for a multitude of duties in rather hard service, the "Model 203" was somewhat larger and a good deal heavier than the average civilian-type training airplane. Placed into service early in 1929, several more examples were soon to follow and duties kept the "203" in service at the school for many years to come. During this time the "203" series underwent several modifications and several conversions to up-grade its performance and utility to meet the greater

demands imposed on an airplane by new methods and new requirements in flight training.

As pictured here, the model "203" series was an open cockpit biplane seating two in tandem; though normally used as a 2-seater, the front cockpit was actually large enough to carry two. Some accounts have labeled this particular Boeing model as a 2-3 place sport craft, suitable for pilot training. As first introduced the model "203" was powered with the 7 cyl. Axelson model A engine of 115 h.p. and following examples were powered with the Axelson B of 150 h.p.; 5 of the "203" were built on a Group 2 approval numbered 2-139. All 5 of these ships were later modified with the installation of the 5 cyl. Wright J6 (R-540-A) engine of 165 h.p. into model "203-A" which were approved by ATC # 211; approval 2-139 was thereby cancelled because all of the "Model 203" were converted to the model "203-A". The type certificate number for the model "203-A" was issued 8-24-29 and altogether some 7 examples of this model were

Fig. 35. Boeing 203 was earlier version with Axelson B engine.

built by the Boeing Airplane Co. of Seattle, Wash. Boeing's engineering staff now consisted of more than 80 men, headed by C. N. Monteith; they had 58 projects going involving 69 aircraft of 6 different types. We might also mention here that in February of 1929, the Hamilton Metalplane Co., builders of the H-45 and H-47 all-metal monoplanes, became a division of the Boeing Company.

Listed below are specifications and performance data for the Boeing model "203-A" as powered with the 165 h.p. Wright J6 engine; length overall 24′4″; hite overall (tail up) 9′11″; wing span upper 34′0″; wing span lower 28′8″; wing chord upper 70″; wing chord lower 54″; total wing area 300 sq.ft.; airfoil Boeing; wt. empty 1789; useful load 788; payload with 40 gal. fuel was 308 lbs.; gross wt. 2577 lbs.; max. speed 110; cruising speed 92; landing speed 50; climb 600 ft. first min. at sea level; ceiling 13,000 ft.; gas cap. 40 gal.; oil cap. 5 gal.; cruising range at 9 gal. per hour was 360 miles; range at 60% power was slightly over 400 miles.

The fuselage framework was built up of welded chrome-moly steel tubing, faired to

Fig. 36. 203-A trainer shows kin to Boeing 95 and 100.

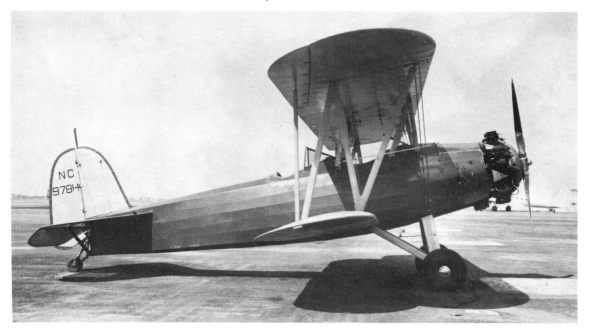

Fig. 37. Used at Boeing flying school, 203-A underwent occasional modification.

shape with wood fairing strips and fabric covered. The cockpits were deep and well protected; bucket seats had wells for parachute pack and dual controls were provided. The wing framework was built up of solid spruce spar beams that were routed out for lightness, with spruce and plywood truss-type wing ribs; the leading edges were covered with plywood to preserve the airfoil form and the completed framework was covered in fabric. The gravity-feed fuel tank was mounted in the center-section panel of the upper wing; Friese type balanced-hinge ailerons were mounted in the upper wing and operated through a streamlined push-pull strut. Large amount of stagger between wing panels, coupled with 67 inch wing gap, offered good visibility during take-offs and landings. The wide tread landing gear was of the cross-axle type and used Boeing oleo shock absorber struts; wheels were 30x5 and Bendix brakes were standard equipment. The "203-A" trainer used either tail-skid or tail-wheel, and some later conversions used low-pressure "air-wheels". The fabric covered tail-group was built up of welded steel tubing; the fin was ground adjustable and the horizontal stabilizer was adjustable in flight. A metal propeller, navigation lights, and hand-crank inertia-type engine starter were standard equipment. Various modifications performed on this series are clearly visible in the various views presented. The next Boeing development was the low-winged "Monomail" trans-

port that will be discussed in the chapter for ATC # 330.

Listed below are Boeing "203-A" entries that were gleaned from various registration records:

    ; Model 203-A (# 1137) Wright J6-5-165
NC-587K ; Model 203-A (# 1138)   „
NC-977H ; „   „ (# 1139) „
NC-978H ; „   „ (# 1140) „
NC-979H ; „   „ (# 1141) „
NC-12748; „   „ (# 1940) „
NC-13392; „   „ (# 1986) „

Registration number for serial # 1137 unknown; first five airplanes were originally as model 203 with Axelson engine, later converted to 203-A with Wright J6-5-165; last two airplanes were first as 203-A, later converted to 203-B with 215 h.p. Lycoming engine; model 203-B was on Group 2 approval numbered 2-412 for serial # 1138 1139-1940-1986.

Fig. 38. Last version was 203-B with Lycoming engine; forerunner to famous Boeing "Kaydet."

Fig. 39. SM-1FS with 300 H.P. Wright J6 engine was SM-1F on EDO pontoons.

The model SM-1FS was a pontoon-mounted version of the popular 6 place Stinson "Detroiter" cabin monoplane, as powered with the 9 cyl. Wright J6 engine of 300 h.p. Basically similiar to the wheel-equipped model SM-1F (see chapter for ATC # 136 of U.S. CIVIL AIRCRAFT, Vol. 2), the SM-1FS mounted either Edo model J or model K float-gear and performed admirably in this configuration. At least one was known to operate very well and profitably too in Alaskan territory near Ketchikan; it was performing the usual north-country services of hauling passengers, fur pelts, machinery, supplies, and whatever, either separately or sometimes all together. Though perhaps overshadowed by the popularity of other Stinson monoplanes of this period, especially the popular "Junior", the trust-worthy SM-1F "Detroiter" sold quite well and remained in limited production into 1931; quite a few were still in active service some 10 years later.

The float-mounted model ≠ SM-1FS as powered with the 300 h.p. Wright J6 engine, received its type certificate number on 8-24-29. Records do not show how many examples of this model were manufactured by the Stinson Aircraft Corp. of Wayne, Mich., but serial numbers A-509 and upwards were eligible for modification to the seaplane specification. Directors of Stinson Aircraft Corp.

also felt the pinch in aircraft sales caused by the economic depression of late 1929, so they accepted an offer by the huge Cord Corp. to purchase control; this was strictly a move for sounder finance during these troubled times. E. L. Cord was chairman of the board; Edward A. Stinson was president; Harvey J. Campbell was V.P.: Wm. A. Mara was secretary; Kenneth M. Ronan and W. C. Naylor were project engineers; Bruce A. Braun who had been chief pilot at Stinson for several years now, was promoted to factory superintendent.

Listed below are specifications and performance data for the float-equipped Stinson "Detroiter" model SM-1FS as powered with the 300 h.p. Wright J6 engine and mounted on Edo model K metal pontoons; length overall 34'9"; hite overall 10'2"; wing span 47'1"; wing chord 84"; total wing area 292 sq.ft.; airfoil modified M-6; wt. empty 3198; useful load 1502; payload with 91 gal. fuel was 730 lbs.; gross wt. 4700 lbs.; (the installation of Edo model J floats held useful load and gross wt. to 100 lbs. less than figures shown above); max. speed 120; cruising speed 100; landing speed 65; climb was 520 ft. first min. at sea level; ceiling 14,000 ft.; gas cap. 91 gal.; oil cap. 7 gal.; cruising range at 15 gal. per hour was 550 miles; price at the factory was

Fig. 40. SM-1F that operated part of year on floats in Canada.

$13,500. plus the float installation, for an average close to $15,000. In mid-1930 the SM-1F was offered with the 300 h.p. Pratt & Whitney "Wasp Jr." engine for $11,245; installation of Edo float-gear was extra.

In most all instances, construction details were identical or basically similiar between the models SM-1F and SM-1FS; the fuselage of the seaplane was plainly marked "SM-1FS", indicating that it had been properly strengthened in appropriate places for the mounting of the float-gear. A full load in the SM-1FS occupying all 6 places allowed 150 lbs. for the pilot and 150 lbs. for each of the 5 passengers; no baggage was allowed. The alternative for more payload was to cut down on the fuel load, which in this case was often held to 70 gal.; cruising range was only cut by some 100 miles. Stinson SM-1F serial numbers of A-509 and upwards were eligible for modification to SM-1FS specifications. The next development in the Stinson monoplane was the 8 place model SM-6B as discussed in the chapter for ATC # 217 in this volume.

A.T.C. #213
(8-26-29)
CURTISS "FALCON", CARGO-PLANE

Fig. 41. Curtiss "Falcon Mailplane" with 600 H.P. geared "Conqueror" engine.

Some airplanes have that certain carriage and look of "romance" that just seems to radiate from them in any angle, and the distinctive swept-wing "Falcon" cargo-plane was surely one of these; its jaunty stance and rugged profile seemed to convey to one a feeling of assured success in man's battle with chance and the elements to perform a difficult task. Though well removed from its infancy, scheduled air-mail service was yet a rather difficult task and not yet aided by the new inventions designed to lighten the burdens of this chore; so it yet required a vehicle that could stand up to the dangers that the elements and the terrain had to offer, and help "the man at the stick" who had other aids too, but still relied mostly on accumulated experience, fortitude, and a measure of good luck.

Developed from the Curtiss "Falcon" O-1, a very illustrious breed of airplane that was used by the Air Corps as an observation aircraft, the mail-carrying "Falcon" was already in active service on extensive routes with N.A.T.; these were "Liberty 12" powered versions that were discussed in Vol. 2 of U.S. CIVIL AIRCRAFT. Several early versions were powered with the 12 cyl. vee-type Curtiss

D-12 engine and this latest development of the cargo-carrying "Falcon" was powered with the 12 cyl. vee-type Curtiss "Conqueror" engine of 600 h.p. Cargo capacity was available for a sizeable payload and ample enough even for bulky loads; the cargo was well protected in metal-lined compartments that were fire-resistant and sealed against the weather. Performance of this "Conqueror" powered "Falcon" was indeed commendable under any condition and its geared-down (2 to 1) engine allowed use of a large diameter 2 or 3-bladed propeller for more efficient thrust. Normally this "Falcon" version was a one-place airplane with the pilot placed in an open cockpit far aft in the fuselage, but it was listed also as a 2 place airplane. The second occupant (a daring air-traveller, no doubt) would be seated in the rear cargo compartment just ahead of the pilot's cockpit; in this case a seat would be installed and windows were provided in the hatch cover for vision out. One of the illustrations shown here, pictures this arrangement on an earlier version of the "Falcon" cargo-plane. The "Falcon" cargo-plane as powered with the 12 cyl. geared-down Curtiss "Conqueror" engine of 600 h.p., was awarded a type certificate number on 8-26-29 and only one example of this particular model was manufactured by the Curtiss Aeroplane & Motor Co. at their huge Buffalo, New York plant.

Listed below are specifications and performance data for the cargo-carrying Curtiss "Falcon" as powered with the 600 h.p. "Conqueror" engine; length overall 28'3"; hite overall 10'4"; wing span upper 38'0"; wing span lower 35'0"; wing chord upper 70"; wing chord lower 54"; total wing area 352.7 sq.ft.; airfoil Clark Y; wt. empty 3367; useful load 1898; payload with 135 gal. fuel was 805 lbs.; gross wt. 5265 lbs.; max. speed 160; cruising speed 136; landing speed 66; climb 1400 ft. first min. at sea level; climb in 10 min. was 10,000 ft.; ceiling 18,000 ft.; gas cap. normal 135 gal.; oil capacity 13 gal.; cruising range at 36 gal. per hour was 475 miles.

The fuselage framework was built up of riveted duralumin tubing, with main fittings at various highly-stressed points of welded chrome-moly steel sheet stock and tubing; the framework was faired to shape with fairing strips and fabric covered. The upper section of the fuselage was covered with sheet aluminum panels clear back to the tail-post; the forward panels were hinged to gain access to the cargo compartments and a baggage compartment was located aft to the pilot's seat. The wing panel framework was built up of heavy-sectioned spruce box-type spar beams, with spruce and plywood girder-type wing ribs; the leading edges were covered with dural sheet and the completed framework was fabric covered. Ailerons of the Friese balanced-hinge type were attached to both upper and lower panels, and connected together with a streamlined push-pull strut. The upper wings were swept back from the center-section panel to an angle of 9 degrees, and wing stagger was 43.5 in. at the fuselage. The pilot's cockpit was roomy and well protected; cockpit heat was provided by a heater-muff around the exhaust pipe. The fabric covered tail-group was built up of riveted dural tubing and channel sections; both rudder and elevators were aerodynamically balanced, and the horizontal stabilizer was adjustable in flight. The fuel load was carried in a fuselage tank, a large belly-tank, and a gravity-feed tank mounted in the center-section panel of the upper wing; the belly-tank had a large dump valve to dispose of the fuel in case of a forced landing. The split-axle landing gear used Curtiss oleo and rubber shock absorbing struts; wheels were 36x8 and Bendix brakes were standard equipment. A Curtiss-Reed metal propeller, a hand-crank inertia-type engine starter, oxygen equipment for the pilot, and wiring for a complete complement of night-flying equipment were provided. The geared-down "Conqueror" developed 600 h.p. at 2400 r.p.m.; cooling of this vee-type liquid cooled

Fig. 42. Cargo holds in "Falcon" ample for 800 lbs. of mail.

Fig. 43. "Falcon" was veteran on night routes of N.A.T.

Fig. 44. "Falcon" version for mail and 1 passenger.

engine was with "Prestone", which allowed the engine to operate at higher temperatures for more efficient power. Because of the higher operating temperatures, the amount of cooling liquid needed was far less and so was the radiator core area, a saving in weight and also parasitic resistance. The next mail-carrying development by Curtiss was the "Carrier Pigeon" which is discussed in this volume in the chapter for ATC # 237.

Listed below is the only known example of the "Conqueror-Falcon" as discussed in this chapter:

NC-301E; Conqueror-Falcon (# 15) Geared-Conqueror 600.

Serial # 15 was operated by National Air Transport on various night runs.

Fig. 45. 5C3-B with Axelson A engine was rare version of popular Command-Aire biplane.

It has long been known throughout the industry that manufacturers of aircraft engines would sometime approach an airplane manufacturer with the proposition of furnishing one of their engines for the prototype of a proposed model that was being built up, or perhaps for another version of an already established series. Word got about that this was the agreement that more or less prompted the introduction of the Axelson-powered Command-Aire model 5C3-B; the Axelson, which was a development from the earlier "Floco" engine, was beginning to create an increasing country-wide appeal and it was sincerely felt that this new version would be a worthy addition to the growing list of models in the "Command-Aire" biplane line.

The model 5C3-B, an added entry in the popular 5C3 series (refer to ATC # 184-185 in U.S. CIVIL AIRCRAFT, Vol. 2), was also a three place open cockpit biplane of the all-purpose type, and was typical in most all respects except for its powerplant installation

which in this case was the 7 cyl. radial air-cooled Axelson engine of 115-150 h.p. Inherent performance and flight characteristics were naturally retained and this model shared all of the desirable qualities that made the Command-Aire biplane a long-time favorite. This rare and seldom seen version would have no doubt found particular favor in a certain segment of the aircraft market, had it not been nipped in the bud by the general slump in aircraft sales by year-end of 1929. The model 5C3-B as powered with the Axelson engine, received its type certificate number on 8-26-29 for serial numbers W-61 and up, and some 3 or more examples of this model were manufactured by Command-Aire, Inc. at Little Rock, Arkansas. A Group 2 approval numbered 2-249 was awarded 8-7-30 to a later version of the model 5C3-B.

Listed below are specifications and performance data for the Command-Aire model 5C3-B as powered with the Axelson engine rated at 150 h.p.: length overall 24'4"; hite overall 8'4"; wing span both 31'5"; wing chord both 60"; wing area upper 157 sq.ft.; wing area lower 146 sq.ft. total wing area 303 sq.ft.; airfoil "Aeromarine 2A"; wt. empty 1503; useful load 869; payload with 51 gal. fuel was 358 lbs.; gross wt. 2372 lbs.; max. speed 115; cruising speed 98; landing speed 40; climb 790 ft. first min. at sea level; ceiling 13,500 ft. gas cap. 51 gal.; oil cap. 5 gal.; cruising range at 7.5 gal. per hour was 600 miles; price at the factory was $6125., later reduced to $5250.

The fuselage framework was built up of welded chrome-moly steel tubing, faired to shape with fairing strips and fabric covered. A one-piece sheet metal turtle-back extended from the rear cockpit to the tail-post; cockpit coamings were also sheet metal panels so consequently, less than half of the fuselage framework was covered in fabric. The fuel tank of 51 gal. capacity was mounted high in the fuselage, just ahead of the front cockpit; a baggage compartment of 4 cubic foot capacity was located behind the rear cockpit with an allowance for 15 lbs. The wing framework was built up of solid spruce spar beams, with spruce and plywood truss-type wing ribs; the completed panels were covered with fabric. The upper and lower wings were in two panels each; ailerons of the slotted-hinge type were mounted in the lower panels only. The split-axle landing gear had a tread of 84 inches, using two spools of rubber shock-cord to absorb the bumps; wheels were 28x4 but Bendix 30x5 wheels with brakes were available. A Hartzell wooden propeller was normally used but a metal propeller was available; a hand crank inertia-type engine starter and navigation lights were available as optional equipment. The next development in the Command-Aire biplane series was the Wright-powered model 5C3-C as discussed in the chapter for ATC # 233 in this volume.

Listed below are Command-Aire model 5C3-B entries that were gleaned from registration records:

NC-608;	5C3-B	(#W-61)	Axelson
NC-948E;	„	(#W-94)	„
NC-973E;	„	(#W-111)	„
NC-10457;	„	(# W-142)	„

Serial # W-142 (NC-10457) was built on Group 2 approval numbered 2-249 with allowance for 2510 lbs. gross weight.

Fig. 46. Fairchild KR-21 with 100 H.P. Kinner K5 engine.

Apparently fired into formative action by the almost instant popularity of the "Fleet", Great Lakes 2-T-1, the Alliance "Argo", and other such sport-trainer designs, Kreider-Reisner quickly laid plans for the production of their model C-6 sport-trainer, which was to be eventually designated as the Fairchild KR-21. The model KR-21 was a handsome little biplane with robust proportions and a sure-footed stance that was somewhat apart from the other training plane designs, and offered many features that were particularly adapted to this type of aircraft. Knowing full well that this type of airplane would often be subjected to more than the average abuse, and the forte of its character would certainly be tested time and again by over-enthusiasm at the "control stick", it was therefore important above all that this little sport-trainer be designed and built especially rugged to stand up to this sort of treatment without wavering. The tapered wings of heavy section were designed for all sorts of abnormal loadings, and the rest of this craft was robustly built in proportion; being doubly braced and amply stressed for all manner of air-borne gyrations called "aerobatics". It is known fact that the KR-21 sport-trainer biplane could be "wrung

out" with great enthusiasm, whether by fledgling or expert, and not one showed any signs of weakness in structure or in character. Beside being designed and built stout of frame and of heart, the KR-21 was eager and sensitive but not tricky, and quite a satisfying pleasure to fly; its performance in all other aspects was commendable and certainly comparable to the best. Of the number built through 1930-1931, some 30 examples of the KR-21 were still in active operation some 10 years later, and 2 or 3 at least, were known to be flying in the 1960 period.

Fairly typical to most other craft of this type, the Fairchild model KR-21 was a light open cockpit biplane seating 2 in tandem, with a flexible personality well adapted to the demands of pilot-training or the effervescent traits desired in sport-type flying. For all production versions the standard powerplant was the popular 5 cyl. Kinner K5 engine of 100 h.p. First introduced in late 1928 as the prototype for the KR-21 sport-trainer series, the Kreider-Reisner model C-6 as shown here, was powered with the 7 cyl. Warner "Scarab" engine of 110 h.p.; performance in this combination was very good, but corporate

Fig. 47. Warner powered C-6 was Kreider-Reisner prototype for Kr-21.

obligations decided upon the use of the Kinner K5 engine instead. The robust wing cellule had wing panels that were tapered in plan-form and in section, a large interplane gap was used to lessen the interference, and the small amount of interplane stagger and placement of variable loads near the C.G., caused less need for trim under light loading or full-load operating conditions. Possessing a character that ranged from docile to exhuberant, the KR-21 adjusted itself very well to the amount of experience shown at the "stick", and therefore was perfectly suitable for both primary and advanced training stages. Offered as a stripped-down training craft, minus the superfluous fineries, the KR-21 was also available as a deluxe "Sportster" with all the plush and extra features that would be desirable in a sport-type airplane. The Fairchild model KR-21A (Kreider-Reisner C-6-B) as powered with the 100 h.p. Kinner K5 engine, received its approval for a type certificate number on 8-26-29 and some 45 or more examples of this model were manu-

factured by the Kreider-Reisner Aircraft Co., Inc. at Hagerstown, Maryland; a subsidiary of the Fairchild Airplane Mfg. Corp. John S. Squires was the president; and Louis E. Reisner was V.P.

Listed below are specifications and performance data for the Fairchild model KR-21A as powered with the 100 h.p. Kinner K5 engine; length overall 22'1"; hite overall 8'6"; wing span upper 27'0"; wing span lower 24'6"; wing chord both 57" at root, and 41" at tip; total wing area 193 sq. ft.; airfoil U.S.A. 45 Mod.; wt. empty 1068; useful load 535; payload with 23 gal. fuel was 228 lbs.; gross wt. 1603 lbs.; max. speed 110; cruising speed 95; landing speed 45; climb 775 ft. first min. at sea level; ceiling 12,400 ft.; gas cap. 23 gal.; oil cap. 3 gal.; cruising range at 6 gal. per hour was 330 miles; price at the factory was first quoted at $4685., lowered to $4375., and lowered to $4125. in late 1930. Early versions of the Kreider-Reisner model C-6-B weighed in a little lighter, as follows; wt. empty 1015; useful load 535; payload 228; gross wt. at 1550 lbs.; as a consequence, performance was a bit more sprightly. A later ammendment allowed loadings as follows; 40 lbs. baggage and 22 gal. fuel with gross wt. of 1635 lbs. for trainer; 56 lbs. baggage and 30 gal. fuel with gross wt. of 1700 lbs. for Sportster; performance would be affected slightly in direct proportion.

Fig. 48. KR-21 shown was Kinner Motors demonstrator.

The fuselage framework was built up of

welded chrome-moly steel tubing and faired to shape with wooden fairing strips; the completed framework was fabric covered. The cockpits were deep, offering good weather protection and bucket seats had wells for seat-type parachute packs; engine instruments were provided in both cockpits. The baggage compartment was of 1.5 cu. ft. capacity for a total allowance of 40 to 56 lbs. which included two parachutes at 20 lbs. each. The wing panels which were tapered both in chord and in section, were built up of heavy spruce spar beams that were routed out for lightness, and girder-type wing ribs were built up of spruce diagonals and plywood gussets; the completed framework was covered in fabric. The 4 narrow-chord ailerons were connected together in pairs by a streamlined push-pull strut; the gravity-feed fuel tank was placed in the center-section panel of the upper wing. The split-axle landing gear of 66 inch tread used rubber rings in compression as shock absorbers but this was later changed to oleo-spring shock absorbing struts; wheels were 24x4 and wheel brakes were available as extra equipment. The fabric covered tail-group was built up of welded chrome-moly steel tubing; the fin was ground adjustable and the horizontal stabilizer was adjustable in flight. Standard equipment for the KR-21A included dual controls, engine and cockpit covers, tool kit and fire extinguisher, log books, first-aid kit, booster magneto, wooden propeller, and wiring for navigation lights. Extra equipment available on the "Sportster" version included metal propeller, wheel brakes, navigation lights, engine starter, extra instruments, and custom color designs. The next development in this sport or training biplane was the model KR-21B that will be discussed in the chapter for ATC # 363; the next development in the Fairchild monoplane series was the Model 42 as discussed in the chapter for ATC # 242 in this volume.

Listed below are Fairchild model KR 21A entries that were gleaned from registration records:

Fig. 49. Rugged nature of KR-21 ideal for sport flying.

NC-859H;	C-6-B	(# 306)	Kinner K5
X-860H;	,,	(#)	,,
X-207V;	KR-21A	(# 1)	,,
X-107M;	,,	(# 1011)	,,
NC-108M;	,,	(# 1012)	,,
NC-359N;	,,	(# 1013)	,,
NC-110M;	,,	(# 1014)	,,
NC-109M;	,,	(# 1015)	,,
NC-360N;	,,	(# 1016)	,,
NC-362N;	,,	(# 1017)	,,
NC-363N;	,,	(# 1018)	,,
NC-365N;	,,	(# 1019)	,,
NC-364N;	,,	(# 1020)	,,
NC-367N;	,,	(# 1022)	,,
NC-366N;	,,	(# 1023)	,,
NC-368N;	,,	(# 1024)	,,
NC-203V;	,,	(# 1025)	,,
NC-204V;	,,	(# 1026)	,,
NC-201V;	,,	(# 1027)	,,
NC-202V;	,,	(# 1028)	,,
NC-205V;	,,	(# 1029)	,,
NC-206V;	,,	(# 1030)	,,
NC-208V;	,,	(# 1031)	,,
NC-209V;	,,	(# 1032)	,,

Serial numbers for the KR-21A continued on to # 1053; serial # 254X was prototype of C-6 series; serial number unknown for X-860H; no record of serial numbers 1001 thru 1010; serial # 1011 was sales promotion demonstrator for Kinner Airplane & Motor Co.; registration numbers for serial # 1014-1021-1027-1031-1052 unverified; serial # 1033 tested with Genet "Major" engine; serial # 1053 later modified as the prototype for KR-21B series.

A.T.C. #216
(8-28-29)
NEW STANDARD, D-29A

Fig. 50. New Standard D-29A with 100 H.P. Kinner K5 engine; very popular with student pilots.

The New Standard model D-29-A biplane was a progressive development in the training plane series that were started off with the Cirrus-powered model D-29 (refer to ATC # 198 of U.S. CIVIL AIRCRAFT, Vol. 2). Not entirely satisfactory and somewhat under-powered, the earlier D-29 design was modified here and there, and power was boosted by the installation of the 100 h.p. Kinner K5 engine. These modifications and the boost in power had transformed the model D-29A into a very satisfactory trainer with a sprightly nature, and a durability that was proven in years of hard service. Though still an "ugly duckling" by way of looks, the D-29A was very amiable in nature and possessed of strict obedience, with a good "feel" that seemed to get through to even the most ham-handed student. When spurred on by a good firm hand, the D-29A was maneuverable enough to perform all sorts of "acrobatics", and the structure was rugged enough even to withstand the abnormal strains of "outside loops". As with the previous D-29, simplicity in manufacture, ease of maintenance and ease of repair were taken into primary consideration; these requisites governed the configuration to a large degree,

and had much to do with the final appearance of this airplane.

Introduced late in 1929, the New Standard model D-29A was an open cockpit biplane seating two in tandem, in one elongated cockpit that separated the two occupants only by a windshield and a small instrument panel; an arrangement that came to be called a "bathtub cockpit". Visibility was excellent in all directions to both occupants, and the rugged framework of the entire airplane was strong enough to soak up a good amount of abuse. At least one example of this model has withstood the cruel ravages of time, has been rebuilt and is flying yet today. Powered with a 5 cyl. Kinner K5 engine of 100 h.p., there was sufficient power reserve to ease one out of a tough situation now and then; instructor-pilots enjoyed working in this airplane and had naught but praise for this craft. The type certificate number for the model D-29A as powered with the Kinner K5 engine, was issued on 8-28-29 and some 30 or more examples of this model were manufactured by the New Standard Aircraft Corp. at Paterson, New Jersey. Charles L. Auger, Jr. was the president; Louis G. Randall was general manager; Charles Healy Day, who had many fine aircraft designs to his credit already, was V.P. and chief of engineering; R. S. Komarnitsky was head of the engineering department and assistant to Charles Day.

Listed below are specifications and performance data for the New Standard trainer model D-29A as powered with the 100 h.p. Kinner K5 engine; length overall 24'8"; hite overall 9'0"; wing span upper & lower 30'0"; wing chord both 54"; wing area upper 126 sq. ft.; wing area lower 122 sq. ft.; total wing area 248 sq. ft.; airfoil Clark Y; wt. empty 1165; useful load 625; payload with 22 gal. fuel was 239 lbs.; useful load allowed 190 lbs. for each occupant, 40 lbs. for parachutes, and 50 lbs. of baggage which included 20 lb. tool kit; gross wt. 1790 lbs.; max. speed 98; cruising speed 80; landing speed 42; climb 750 ft. first min. at sea level; ceiling 11,000 ft.; gas cap. 22 gal.; oil cap. 3 gal.; cruising range at 6.5 gal. per hour was 250 miles; price at the factory was $4475., and reduced to $3131. in July of 1930. Earliest version of the D-29A was considerably lighter as shown by figures given here; wt. empty 1075; useful load 535; payload 195; gross wt. 1610 lbs.; performance figures would be improved proportionally.

The fuselage framework was built up of open section duralumin members that were riveted and bolted together into a robust and simple structure that could be easily repaired even in the field with ordinary hand tools; the framework was lightly faired to shape with wood fairing strips and fabric covered. The cockpit was one elongated opening but was separated by an instrument panel and a windshield; two sets of instruments were provided. Bucket type seats had deep wells for seat-type parachute pack, and controls were dual of the joy-stick type; a locker of 2.5 cubic foot capacity provided allowance for 50 lbs. of baggage, which included a 20 lb. tool kit, when carried. The wing framework was built up of laminated spruce spar beams with bass-wood and plywood built-up wing ribs; the leading edges were covered with plywood veneer and the completed framework was fabric covered. The four ailerons were connected together in pairs by a streamlined push-pull strut; the ailerons were a riveted framework of open section dural members and fabric covered. The gravity-feed fuel tank was mounted in the center-section panel of the upper wing and was actually shaped like a deep airfoil section; the upper and lower wing panels were identical and could be interchanged.

Fig. 51. D-29A was "Ugly Duckling" with very amiable nature.

Fig. 52. NT-1 was Navy version of D-29A.

All interplane struts were of streamlined duralumin tubing, and interplane bracing was heavy streamlined steel tie-rods. The fabric covered tail-group was built up of riveted members of open-sectioned duralumin; the fin was ground adjustable and the horizontal stabilizer was adjustable in flight. The split-axle landing gear of 78 inch tread was of chrome-moly steel tubing in the rugged out-rigger type and used oil-draulic shock absorbing struts; wheels were 24x4 or 26x4 and no brakes were provided. A metal propeller, navigation lights, and Heywood air-operated engine starter were optional equipment. With only slight modification, the D-29A was ordered by the Navy as the NT-1 trainer. The next development in the New Standard biplane was the model D-25A which is discussed in the chapter for ATC # 224 of this volume.

Listed below are New Standard model D-29A entries that were gleaned from various registration records:

NC-9195; D-29A (# 1001) Kinner K5.
NC-36K;   ,,   (# 1002)   ,,
NC-151M;   ,,   (# 1003)   ,,
NC-152M;   ,,   (# 1004)   ,,
NC-153M;   ,,   (# 1005)   ,,

NC-154M;   ,,   (# 1006)   ,,
NC-155M;   ,,   (# 1007)   ,,
NC-156M;   ,,   (# 1008)   ,,
NC-157M;   ,,   (# 1009)   ,,
NC-158M;   ,,   (# 1010)   ,,
NC-164M;   ,,   (# 1011)   ,,
NC-165M;   ,,   (# 1012)   ,,
NC-166M;   ,,   (# 1013)   ,,
NC-167M;   ,,   (# 1014)   ,,
NC-168M;   ,,   (# 1015)   ,,
NC-169M;   ,,   (# 1023)   ,,
NC-170M;   ,,   (# 1024)   ,,
NC-171M;   ,,   (# 1025)   ,,
NC-922V;   ,,   (# 1027)   ,,
NC-923V;   ,,   (# 1028)   ,,
NC-924V;   ,,   (# 1029)   ,,
NC-925V;   ,,   (# 1030)   ,,
NC-926V;   ,,   (# 1031)   ,,
NC-913V;   ,,   (# 1032)   ,,
NC-716Y;   ,,   (# 1034)   ,,

Registration numbers were not available for serial # 1016-1017-1018-1019-1021 and 1022; serial # 1020 modified into D-29S on Group 2 approval numbered 2-272; serial # 1024 later had Walter engine in Mexico; serial # 1026 as D-29 Spl. with Menasco "Pirate" B-4; serial # 1006-1025-1033 as model D-31 on Group 2 approval numbered 2-276.

A.T.C. #217
(8-28-29)
STINSON "DETROITER", SM-6B

Fig. 53. Model SM-6B was big "Detroiter" with 420 H.P. Wasp engine.

The biggest airplane in the popular Stinson "Detroiter" monoplane line was the model SM-6B as shown here; a large high wing cabin monoplane with seating for 8, or the ability to carry better than 950 pounds of cargo. Looking very much like an enlarged "Detroiter" SM-1F, the model SM-6B was especially designed for the smaller air-lines that offered shuttle-service on feeder-routes connecting with the main transcontinental system. With passenger seating being quickly removable, the SM-6B was ideal too for hauling air-freight and bulky cargo in a cabin of 48x49x92 inch dimensions; its durability and utility in the "bush country" drew frequent praise. Though quite rare in number built, several examples of the buxom SM-6B served for many years in out-of-the-way places of this country. Eddie Stinson was sincerely proud of his "Detroiter" monoplanes; they say he was especially proud of this one.

The big SM-6B was a high wing cabin monoplane seating 8 with ample room and good comfort; powered with a Pratt & Whitney "Wasp" C-1 of 420-450 h.p., this craft possessed an agility and lively performance that certainly belied its weight and apparent bulk. Typical to other "Stinson" monoplanes of this memorable period, the big model

SM-6B was not a pampered lady; built rugged with inherent personality characteristics typical of all "Detroiter" monoplanes, this 8 place version was a worthy addition to the popular line. Tested extensively in the first part of 1929, the first two examples of the SM-6B were awarded approval as 7 place craft at 5000 lb. gross; this was a Group 2 approval numbered 2-89 and issued 7-3-29. The type certificate number for the 8 place SM-6B as powered with the 450 h.p. "Wasp" engine was issued 8-28-29 and some 7 or more examples of this model were manufactured by the Stinson Aircraft Corp. at Wayne, Michigan. Edward A. Stinson was president; Harvey J. Campbell was V.P.; William A. Mara was secretary; with Kenneth M. Ronan and W. C. Naylor as project engineers.

Listed below are specifications and performance data for the Stinson "Detroiter" model SM-6B as powered with the 450 h.p. "Wasp" engine; length overall 34'4"; hite overall 9'8"; wing span 52'8"; wing chord 84"; total wing area 334 sq.ft.; airfoil M-6; wt. empty 3496; useful load 1854; payload with 110 gal. fuel was 954 lbs.; gross wt. 5350 lbs.; max. speed 148; cruising speed 128; landing speed 60; climb 1000 ft. first min. at sea level; ceiling 18,000 ft.; gas cap. 110 gal.;

Fib. 54. Veteran SM-6B operated for many years in the bush country.

oil cap. 9.5 gal.; cruising range at 23 gal. per hour was 575 miles; price at the factory field was $18,500., reduced to $18,000., and finally offered for $15,995. in 1930.

The fuselage framework was built up of welded chrome-moly steel tubing and was gusseted with chrome-moly sheet stock at every joint to make up a structure of exceptional strength and rigidity; the completed framework was faired to shape with formers and fairing strips, then fabric covered. The cabin walls were sound-proofed and insulated, with upholstery of mohair fabric in several tasteful combinations; the cabin was heated by hot air coming off an exhaust manifold muff and

ventilation was provided by roll-down windows. All windows were of shatter-proof glass and could be rolled up or down; a baggage compartment was to the rear of the cabin area and was accessible from either inside or outside. No baggage was allowed with 8 passengers and 110 gal. fuel. A large door on each side with a large convenient step was provided for entry and exit. The wing framework was built up of heavy-sectioned solid spruce spars that were routed to an I-beam section; wing ribs were built up of spruce and plywood in a truss-type form. The leading edge was covered with duralumin sheet to preserve the airfoil form, and the completed framework was covered with

Fig. 55. 8 place SM-6B operated well from small air-fields.

fabric. The Friese type ailerons were of a welded steel tube frame work, covered with fabric; the fuel supply was carried in two tanks mounted in the wing, flanking each side of the fuselage. The wing bracing struts were heavy gauge chrome-moly steel tubes that were encased in balsa-wood fairings; these fairings were shaped to an Eiffel 380 airfoil section to provide added lift and stability. The fabric covered tail-group was built up of welded chrome-moly steel channel sections and tubing; the fin was ground adjustable and the horizontal stabilizer was adjustable in flight. The elevators were operated by a push-pull tube and the rudder was operated by cables; dual "dep" wheel controls were provided. The split-axle landing gear of 114 inch tread was of the outrigger type and used air-oil shock absorbing struts; wheel brakes and a steerable tail-wheel made ground maneuvering a fairly easy task. Wheels were 32x6 and Bendix brakes were standard equipment; a metal propeller, navigation lights, battery, and electric inertia-type engine starter were also offered as standard equip-

ment. The next development in the Stinson monoplane was the model SM-8B, a "Junior" powered with the 7 cyl. Wright J6 engine of 225 h.p.; for discussion of this model see the chapter for ATC # 294 in this volume.

Listed below are Stinson model SM-6B entries that were gleaned from various registration records:

NC-9697;	SM-6B	(# 2000)	P & W Wasp
NC-8426;	,,	(# 2001)	,,
;	,,	(# 2002)	,,
;	,,	(# 2003)	,,
NC-460H;	,,	(# 2004)	,,
NC-463H;	,,	(# 2005)	,,
NC-468H;	,,	(# 2006)	,,
NC-484H;	,,	(# 2007)	,,
;	,,	(# 2008)	,,
NC-407M;	,,	(# 2009)	,,
NC-412M;	,,	(# 2010)	,,
;	,,	(# 2011)	,,
NC-219W;	,,	(# 2012)	,,

Registration numbers for serial # 2002-2003-2008-2011 are unknown.

Fig. 56. Monoprep trainer with Velie M-5 engine; lovable airplane was long-time favorite.

It is certainly not sufficient to say that the "Monoprep" trainer was developed from the basic lay-out of the "Monocoupe 113" and was therefore typical, because although it was a sister-ship and resembled it to a great extent, it had a contagious personality that affected one somewhat differently than the "model 113". Of the many that will remember the "Monoprep", or have actually learned to fly in one, most will surely say that it was a thoroughly delightful craft and probably one of their all-time favorites. In the words of one admiring pilot, "It was the doggone-dest airplane you ever sat in, and you just couldn't help but like it". Personality analysis of the "Monoprep" reveals a pleasant nature that was not easily disturbed, a friendliness that seemed to get through to most everybody, and a capability that would adjust itself according to the experience it felt on the controls. If a timid first-ride student, skidding his way through turns and such, the "Prep" would bear with you because inherently it was a good trainer; but its playful nature was also sensitive to a good hand on the "stick" and it was always ready to do as bid. Pilots with thousands of hours still enjoyed taking a turn or two in the "Monoprep".

The pixie-like "Monoprep" was a fitting companion to the loveable "Monocoupe"; arranged in a semi-cabin lay-out with open sides and a large windshield up front, the parasol-winged "Prep" allowed somewhat more room in the cockpit for bulky clothing and allowed the student to feel-the-wind on skidding turns, side-slips, and the like. The "Monoprep" as pictured here, was a high-winged monoplane seating two side by side and it was powered with the 5 cyl. Velie M-5 engine of 55 h.p.; the "parasol" arrangement was not as "clean" in the aerodynamic sense, but performance was nearly comparable to that of the " Monocoupe". As first introduced on the "Monocoupe" of 1927, the chummy side-by-side seating promoted companionship and easy on-the-spot communication; this was a feature ideal for pilot instruction. As a primary trainer, the "Monoprep" was considered one of the best and it had enough inherent capability to give a student-pilot a good preview of the things he was expected to learn in his secondary phase. Numerous flying-schools used the "Monoprep" for the primary phase of their pilot-training courses and the students were usually ready and eager to take their turn. A few of the "Monoprep", fitted with 90 or 110 h.p. engines, were used strictly for sport flying.

Introduced in the early part of 1929, the first few examples of the "Monoprep" were fitted with a long-legged landing gear as used on the "Monocoupe 113", but production versions were soon fitted with the short-legged landing gear as used on the "Monosport". Designed to do a job with efficiency and simplicity, the "Monoprep" was easy to operate, easy to fly, easy to get in and out of, and easy to service; the rugged structure could absorb plenty of abuse and hard service so idle-time in the hangar was kept at a minimum. Airplanes used for pilot-training usually lead a hard life so the mortality rate is bound to be a bit higher than ordinarily; it is therefore quite commendable to the series that at least 22 of the "Monoprep" were still being used in active service on the flight-line some 10 years later. The first batch of "Monoprep" to come off the line were certificated on a Group 2 approval numbered 2-90 (issued 8-7-29) for some 8 airplanes; this later amended to cover some 20 airplanes. The type certificate number for the "Monoprep 218" as powered with the 55 h.p. Velie engine was issued 8-30-29 and was eligible for at least 35 airplanes. On 9-10-29 another Group 2 approval, numbered 2-128, was issued for six more airplanes; it is difficult to make a correct tally of the number of "Monoprep" that were manu-factured because accurate records have long been buried, but registration records show that at least 60 or more examples of this model were built by the Mono Aircraft Corp. at Moline, Illinois.

Listed below are specifications and performance data for the "Monoprep 218" as powered with the 55 h.p. Velie M-5 engine; length overall 21'0"; hite overall 6'3"; wing span 32'0"; wing chord 60"; total wing area 143 sq.ft.; airfoil "Clark Y"; wt. empty 783; useful load 505; payload with 15 gal. fuel was 250 lbs.; gross wt. 1288 lbs.; max. speed 92; cruising speed 80; landing speed 37; climb 680 ft. first min. at sea level; ceiling 9000 ft.; gas cap. 15 gal.; oil cap. 1.5 gal.; cruising range at 4 gal. per hour was 290 miles; price at the factory upon introduction was $2675. but soon raised to $2835., and lowered to $2575. in May 1930; both airplane and engine carried a factory guarantee as a unit.

The fuselage was a rectangular framework built up of welded chrome-moly and 1025 steel tubing, faired to shape with wooden formers and fairing then fabric covered. The cockpit interior was plain and simple but of ample proportion to allow room for parachutes and bulky clothing. A sky-light in the roof of the cockpit provided vision overhead and also provided additional head-room when sitting on parachute packs; a convenient step

Fig. 57. "Prep" was flying school favorite.

Fig. 58. Early Monoprep used long-leg landing gear; likened to Monocoupe 113 with open cockpit.

and a full-length door were provided on each side for exit and entry. Baggage allowance was 65 lbs., or 25 lbs. when carrying parachutes; dual controls were of the joy-stick type. The semi-cantilever wing framework, spliced into a one-piece construction, was built up of solid spruce spar beams with spruce and basswood wing ribs; the completed framework was fabric covered. The wing was fastened atop 4 inverted-vees of steel tubing and braced to the fuselage by 2 vee-struts of heavy-gauge steel tubing; the two gravity-feed fuel tanks were mounted in the wing flanking the sky-light opening.

The split-axle landing gear was of the short-legged type and used "Mono-oil" (spring & oil) shock absorbing struts; wheels were 26x4 and no brakes were provided. The fabric covered tail-group was built up of welded 1025 steel tubing. The fin was ground

adjustable and the horizontal stabilizer was adjustable in flight. A spring-leaf tail skid and a Hartzell wooden propeller were standard equipment; wiring for navigation lights and wheel brakes were optional equipment. The next "Mono" development was the high performance "Monosport" as discussed in the chapter for ATC # 249 in this volume.

Listed below are "Monoprep 218" entries as gleaned from registration records:

NC-8986;	Monoprep 218	(# 6057)	Velie M-5
NC-8987;	„	(# 6058)	„
NC-8990;	„	(# 6059)	„
NC-8991;	„	(# 6060)	„
NC-8992;	„	(# 6061)	„
NC-8994;	„	(# 6063)	„
NC-8997;	„	(# 6065)	„
NC-8999;	„	(# 6067)	„
NC-101K;	„	(# 6068)	„
NC-103K;	„	(# 6070)	„
NC-109K;	„	(# 6071)	„
NC-110K;	„	(# 6072)	„
NC-111K;	„	(# 6073)	„
NC-112K;	„	(# 6074)	„
NC-115K;	„	(# 6075)	„
NC-118K;	„	(# 6076)	„
NC-119K;	„	(# 6077)	„
NC-120K;	„	(# 6078)	„
NC-122K;	„	(# 6079)	„
NC-123K;	„	(# 6080)	„

Fig. 59. Monoprep arrangement ideal for pilot training.

NC-124K;   „   (# 6081)   „
NC-126K;   „   (# 6082)   „
NC-127K;   „   (# 6083)   „
NC-128K;   „   (# 6084)   „
NC-129K;   „   (# 6085)   „
NC-130K;   „   (# 6086)   „
NC-131K;   „   (# 6087)   „
NC-133K;   „   (# 6088)   „

NC-137K;   „   (# 6089)   „
NC-138K;   „   (# 6090)   „

Serial # 6000 - 6001- 6002 - 6003 were experimental prototypes not covered by approval; Group 2 approval 2-90 issued 8-7-29 for serial # 6004 - 6005 - 6006 - 6013 - 6014 - 6015 - 6016 - 6017 - 6050 and up with Velie 55 at 1360 lbs. gross; same approval amended 9-10-29 for serial # 6004 - 6005 - 6006 - 6013 - 6014 - 6015 - 6016 - 6017 - 6021 - 6025 to 6035 with Velie 65 at 1360 lbs. gross; same approval amended 10-15-29 for serial # 6004-6005-6006-6013 to 6017-6021-6025 to 6035 with Velie 55 at 1360 lbs. gross; Group 2 approval 2-128 issued 9-10-29 for serial # 6050 to 6055 with Velie 55 at 1288 lbs. gross; no listing appears in registration records for serial # 6007 - 6008 - 6009 - 6010 - 6011 - 6012 6018 - 6019 - 6020 - 6022 - 6023 - 6024 - 6036 - 6037 - 6038 - 6040 - 6041 - 6042 - 6043 - 6044 - 6045 - 6046 - 6047 - 6048 - 6049; registration number for serial # 6056 - 6062 - 6064 - 6066 - 6069 unknown; serial # 6087 also had Lambert 90 engine.

Fig. 60. This "Mystery Monoprep" was early version; note features not typical to standard "Prep."

A.T.C. #219
(9-4-29)
KEYSTONE-LOENING "COMMUTER", K-84

Fig. 61. Keystone-Loening "Commuter" model K-84 with 300 H.P. Wright J6 engine.

Loening was among the first to bring forth a practical and successful amphibious aircraft; introduced in early 1925, this craft certainly surprised one and all with its excellent utility and above-average performance. Put into use with the Army and Navy service, further development of this basic design led directly to the commercial versions of the C2C and C2H "Air Yacht", which enjoyed popularity and extensive use across the nation and abroad in air-ferry operations. The new Loening "Commuter" as shown here in various views, was also an amphibious aircraft that was primarily designed for and especially leveled at the sportsman-pilot and the business man; a market for this type of craft that was just beginning to show some possibilities. Arranged as a comfortable 4 place cabin airplane, this new version was also highly versatile in the fact that it could pick and choose its airports or landing-places to the best advantage or wherever fancy would dictate. Unlike the previous "Air Yachts", which were called the "flying shoe-horn", the "Commuter" was basically a "flying boat" configuration with the engine mounted between the

wings in tractor or puller fashion. The tractor configuration did have some advantages, but it was not as practical as the pusher-engine installation for this type of airplane. However the rugged "Commuter" did enjoy a a good measure of popularity and success, and was used many times for exacting and unusual service; the New York City Police Dept. operated one to patrol the harbor area, and another served in the north country with Alaskan Airways. Of very rugged constitution, several of the K-84 were still in service many years later.

Introduced in early 1929, the Keystone-Loening "Commuter" model K-84 was a biplane of the classic "flying boat" type; a cabin of ample proportions with tasteful and practical appointments for four people, was arranged in the forward section of the all-metal hull, with entrance through a large hatch-way. A retractable undercarriage could be extended for operation on land, or swung up out of the way for landings on water; extension or retraction of the landing gear was accomplished by a manually operated hand-crank accessible to the pilot. A large baggage compartment was provided for luggage and stowage of anchoring gear. The powerplant for the K-84 was the 9 cyl. Wright J6 engine of 300 h.p., which was mounted between the wing panels in a streamlined nacelle. Performance of this model was quite average for a craft of this type and flight characteristics were described as gentle and predictable. The type certificate number for the model K-84 "Commuter" as powered with the 300 h.p. Wright J6 engine was issued 9-4-29 and some 40 examples of this model were manufactured by the Loening Aeronautical Engineering Co. of New York City, which was a division of the Keystone Aircraft Corp. Through affiliations with the Curtiss-Wright Corp., "Commuter" sales and distribution were handled at scattered stations throughout the country. Grover Loening, founder of the Loening Aeronautical Engineering Co. and designer of various amphibious types bearing his name since 1925, sold out his interests to the Keystone Corp. and concentrated his efforts on several new developments. Leroy Grumman, able assistant to Grover Loening for several years, also had left the firm and started out on his own. President of the Keystone Aircraft Corp. was Edgar N. Gott; C. T. Porter was V.P. and chief engineer.

Listed below are specifications and performance data for the Keystone-Loening "Commuter" model K-84 as powered with

the 300 h.p. Wright J6 engine; length overall 32'1"; hite on water 12'6"; hite with landing gear extended 13'6"; wing span upper & lower 40'0"; wing chord both 72"; total wing area 437 sq.ft.; airfoil Loening 10-A; wt. empty 2920 useful load 1230; payload with 70 gal. fuel was 593 lbs.; gross wt. 4150 lbs.; max. speed 112; cruising speed 90; landing speed 45; climb 850 ft. first min. at sea level; climb in 10 min. was 5000 ft.; ceiling 12,000 ft.; gas cap. 70 gal.; oil cap. 6.5 gal.; cruising range at 15 gal. per hour was 400 miles; price at the factory was $16,800. As is usually the case, the prototype "Commuter" was somewhat lighter and a bit smaller than the production version. Data showing difference was as follows; span both 36'0"; wing area 389 sq. ft.; wt. empty 2780; useful load 1220; payload with 70 gal. fuel was 583; gross wt. 4000; max. speed 116; cruising speed 96; landing speed 50; ceiling 9800 ft. The last versions of the model K-84, using improved Wright J6 of 330 h.p. were allowed a gross weight of 4270 lbs.

The hull framework was built up of dural frame members that were bolted together into a rigid and durable structure; outer covering aluminum alloy sheet that was bolted and screwed to the hull framework. The bottom of the hull was covered in 1/16 in. duralumin plate and all seams were water-proofed with strips of impregnated fabric. The wing framework was built up of solid spruce spar-

Fig. 62. Commuter behaved well in water; shown here on "step."

Fig. 63. Utility of "Commuter" amphibian ideal for sport flying.

Fig. 64. With gear lowered, commuter taxis up ramp to unload.

beams with stamped-out metal "Alclad" wing ribs; the completed framework was fabric covered. There were four ailerons connected together in pairs by push-pull struts; interplane struts were of streamlined chrome-moly steel tubing and interplane bracing was of streamlined steel wires. Metal-framed and metal covered wing-tip floats were mounted on outer end of lower wings to keep the craft from heeling over during water operations. The fabric covered tail-group was built up with a combination of wood spars and dural former ribs; the fin was ground adjustable and the horizontal stabilizer was adjustable in flight. The retractable landing gear was fitted with oleo-spring shock absorbing struts; wheels were 30x5 or semi-balloon tires were later available. Baggage compartment was allowed a maximum of 89 lbs., which included 40 lbs. of anchoring gear. A metal propeller, Eclipse electric inertia-type engine starter, fire extinguisher, anchor and rope, dual controls, life preservers, and a bilge pump, were offered as standard equipment. The "Commuter" design was revived some years later as a high wing cabin monoplane with engine mounted on a tripod mount above the wing; the hull was basically the same as the K-84 and engine was also mounted in a tractor drive. The next Keystone-Loening development was the "Cyclone" powered "Air-Yacht" model K-85, which will be discussed in the chapter for ATC # 395. The next Keystone development was the tri-motored "Patrician" model K-78D, which is discussed in the chapter for ATC # 260 in this volume.

Listed below are "Commuter" model K-84 entries that were gleaned from various regis-

tration records:

X-9781;	Model K-84	(# 301)	J6-9-300
NC-60K;	,, ,,	(# 302)	,,
NS-370N;	,, ,,	(# 303)	,,
NC-59K;	,, ,,	(# 304)	,,
NC-63K;	,, ,,	(# 305)	,,
NC-61K;	,, ,,	(# 306)	,,
NC-301V;	,, ,,	(# 311)	,,
NC-374V;	,, ,,	(# 313)	,,
NC-375V;	,, ,,	(# 314)	,,
NC-376V;	,, ,,	(# 315)	,,
NC-535V;	,, ,,	(# 316)	,,
NC-538V;	,, ,,	(# 317)	,,
NC-539V;	,, ,,	(# 318)	,,
NC-540V;	,, ,,	(# 319)	,,
NC-339W;	,, ,,	(# 320)	,,
NC-340W;	,, ,,	(# 321)	,,
NC-10247;	,, ,,	(# 323)	,,
NC-10248;	,, ,,	(# 324)	,,
NC-10249;	,, ,,	(# 325)	,,
NC-10250;	,, ,,	(# 326)	,,
NC-19E;	,, ,,	(# 327)	,,
NC-20E;	,, ,,	(# 328)	,,
NC-21E;	,, ,,	(# 329)	,,
NC-22E;	,, ,,	(# 330)	,,
NC-755W;	,, ,,	(# 331)	,,
NC-756W;	,, ,,	(# 332)	,,
NC-757W;	,, ,,	(# 333)	,,
NC-758W;	,, ,,	(# 334)	,,
NC-762W;	,, ,,	(# 335)	,,
NC-763W;	,, ,,	(# 336)	,,

Serials # 307 - 308 - 309 - 310 - 312 - 322 - 337 - 338 - 339 - 340 unaccounted for in registration records up to Jan. of 1932; serial # 313 later as model K-84W with P & W "Wasp" Junior of 300 h.p. on Group 2 approval numbered 2-526.

Fig. 65. Curtiss-Robertson "Robin" J-1 with 165 H.P. Wright J6 engine.

Designed and built for dependable service, the familiar "Robin "continued on in country-wide popularity and was now offered as the model J-1 with the 5 cyl. Wright J6 engine of 165 h.p. This new version was an immediate success with operators that specialized in the varied chores of general-purpose flight operations; business people were also attracted to its economy and utility. The "Robin" was an easy airplane to maintain and to fly; basically forgiving in nature, it was very friendly and exceptionally dependable in every day service. It could hardly be described as feminine or dainty, but the "Robin" was nevertheless a lady and its personality was of honest simplicity. Surprisingly maneuverable for a ship of this type, the "Robin" was a sight to marvel at while doing "slow rolls"; loops were also performed with comparative ease. Certainly not designed for aerobatics, it was a comfort to know that the "Robin" was capable to perform and withstand the stresses of these maneuvers.

Already famous for the setting of several endurance records, the "Robin" put forth a sort of last effort in 1935 by setting yet another record for refueled in flight endurance; flown by the brothers Fred and Al Keys, their continuous flight was sustained to over 653 hours, which was more than 27 days in the air. Their old "Robin", of the model J-1 type, was powered with the 5 cyl. Wright J6 engine and was named "Ole Miss". Probably the most noted flight for the "Robin" J-1 type was the flight to Ireland in 1938 by Douglas "Wrong Way" Corrigan; a flight that started in New York with a plan to fly back to Calif., but he somehow "miscalculated" on his directions and flew the Atlantic Ocean to Dublin, Ireland instead. Corrigan's "Robin", a many times second-hand airplane, was originally powered with the Curtiss OX-5 engine, but was converted by Corrigan to the model J-1 configuration with a Wright J6-5-165 engine.

The "Robin" model J-1 as shown here, was also a high-wing cabin monoplane with seating for three, had baggage allowance for 50 lbs., and was quite typical to other models built previously, except for the installation of

Fig. 66. "Wrong Way" Corrigan and his rebuilt second-hand "Robin" flew Atlantic Ocean.

the 5 cyl. Wright J6 series engine. From hangar-talk it is suspected that grumblings about certain annoying characteristics of the 6 cyl. Curtiss "Challenger" engine, launched the development of the model J-1 with the Wright J6; had this J-1 version been brought out somewhat earlier, there is no doubt that many more would have been built and sold. The popularity of the "Robin" and its rugged character tended to promote longevity, so it is no wonder that over 300 were still in active operation some 10 years later. The type certificate number for the "Robin" model J-1 as powered with the Wright J6 engine of 165 h.p., was issued 9-5-29 and then amended on 10-21-29 to include seaplane version and increased gross weight allowance to 2523 lbs. for land plane; at least 60 examples of this model were manufactured by the Curtiss-Robertson Airplane Mfg. Co. at Anglum (St. Louis), Mo., a division of the Curtiss-Wright Corp.

Listed below are specifications and performance data for the "Robin" model J-1 as

Fig. 67. 3 place model J-1 was popular "Robin" version.

powered with the 165 h.p. Wright J6 engine; length overall 25′6″; hite overall 8′0″; wing span 41′0″; wing chord 72″; total wing area 223 sq.ft.; airfoil Curtiss C-72; wt. empty 1625; useful load 898; payload with 50 gal. fuel was 393 lbs.; gross wt. 2523 lbs.; max. speed 118; cruising speed 100; landing speed 45; stall speed 53; climb 640 ft. first min. at sea level; climb in 10 min 5750 ft.; ceiling 12,800 ft.; gas cap. 50 gal.; oil cap. 5 gal.; cruising range at 9 gal. per hour was 540 miles; basic price at factory was $7000., lowered to $5995. in June of 1930. Figures for the earliest version of the model J-1 as approved on 9-5-29 are as follows; wt. empty 1542; useful load 898; gross wt. 2440 lbs.; there was a proportionate increase in some performance figures. The deluxe version of the model J-1 weighed in as follows; wt. empty 1683; useful load 898; gross wt. 2581 lbs.; performance differences would hardly be noticeable. The seaplane version of the model J-1, mounted on twin-float gear, weighed in as follows; wt. empty 1790; useful load 870; payload with 50 gal. fuel was 365 lbs.; gross wt. 2660 lbs.; max. speed 110; cruising speed 93; landing speed 47; climb 630 ft. first min. at sea level; ceiling 12,500 ft.; gas cap. 50 gal.; oil cap. 5 gal.; cruising range at 9 gal. per hour was 475 miles.

Construction details and interior arrangements have been discussed on numerous occasions for previous "Robin" models, therefore we suggest a review of the descriptions in the chapters for ATC # 40 - 63 - 68 - 69 - 143 - 144. Deluxe version of the model J-1 contained the following extra equipment; adjustable metal propeller, Eclipse inertia-type engine starter, booster magneto, cabin heater, and storage battery. The next "Robin" development was the "Model W" as discussed in the chapter for ATC # 268 in this volume.

Listed below is a partial listing of the "Robin" model J-1 entries that were gleaned from various registration records:

NC-12H;	Model J-1	(# 382)	J6-5-165	
NC-13H;	„	„	(# 384)	„
NC-984K;	„	„	(# 616)	„
NC-701M;	„	„	(# 640)	„
NC-705M;	„	„	(# 644)	„
NC-745M;	„	„	(# 648)	„
NC-748M;	„	„	(# 651)	„

NC-750M;	„	„	(# 653)	„
NC-752M;	„	„	(# 655)	„
NC-758M;	„	„	(# 661)	„
NC-762M;	„	„	(# 663)	„
NC-764M;	„	„	(# 665)	„
NC-766M;	„	„	(# 667)	„
NC-768M;	„	„	(# 669)	„
NC-770M;	„	„	(# 671)	„
NC-772M;	„	„	(# 673)	„
NC-774M;	„	„	(# 675)	„
NC-776M;	„	„	(# 677)	„
NC-778M;	„	„	(# 679)	„
NC-780M;	„	„	(# 681)	„

Serial # 382 also as model J-2 on ATC # 221; serial # 384 first as Robin W, later modified to J-1; serial # 616 first production version; last production version of the model J-1 appears to be serial # 759.

A.T.C. #221
(9-5-29)
CURTISS-ROBERTSON "ROBIN", J-2

Fig. 68. Curtiss-Robertson "Robin" model J-2 with 165 H.P. Wright J6 engine.

Like the earlier "Robin" model C-2, the model J-2 version was more or less a special-category airplane that was arranged slightly different from the standard production model to fit more varied uses. A substantial increase in fuel capacity for a greater cruising range was among the main differences, and allowable total gross weight was slightly higher. As a photo-plane with camera installed in the fuselage floor, the "Robin" J-2 was no doubt used for mapping and general photographic work. With a cruising range of some 8 hours duration, the operating ability of the model J-2 was extended to more than 800 miles. One could then assume that the J-2 would be ideal as an air-taxi in the bush country, where fueling stops were often-times a good ways apart.

Type certificate specifications state that "Robins" with serial # 382 and upwards were eligible as model J-2 aircraft, but only 2 examples are shown in registration records through 1931. It is entirely possible that more "Robins" were converted to this specification at a later date. The type certificate number for the "Robin" model J-2, as powered with the 5 cyl. Wright J6 (R-540) engine of 165 h.p., was issued 9-5-29 and this version was converted in the shops of the Curtiss-Robert-

son Airplane Mfg. Co. at Anglum (St. Louis), Mo., a division of the Curtiss-Wright Corp. In reference to the model C-2 as described in the chapter for ATC # 144 of U.S. CIVIL AIRCRAFT, Vol. 2, we offer to make the following additions to the listed specifications. Fuel capacity was a maximum of 80 gal.; extra fuel was sometime carried in a gravity-feed tank faired into the top side of fuselage, thereby eliminating sky-light panel in the roof of the cabin; wing-tank capacity could also be increased to accommodate 80 gallons. The "Robin" model C-2 as powered with the Curtiss "Challenger" engine, was also eligible for a camera installation and allowance for 114 lbs. of baggage.

Listed below are specifications and performance data for the "Robin" model J-2 as powered with the 165 h.p. Wright J6 engine; length overall 25'6"; hite overall 8'0"; wing span 41'0"; wing chord 72"; total wing area 223 sq.ft.; airfoil Curtiss C-72; wt. empty 1565; useful load 1035; payload with 78 gal. fuel was 357 lbs.; payload with 50 gal. fuel was 545 lbs; gross wt. 2600 lbs; max. speed 118; cruising speed 100; landing speed 48; stall speed 54; climb 635 ft. first min. at sea level; climb in 10 min. was 5700 ft.; ceiling 12,700 ft.; gas cap. normal 50 gal.; gas cap.

78 gal.; oil cap. 5 gal.; cruising range at 9 gal. per hour was 540-850 miles; basic price at the factory was approx. $6000. "Robin" with serial # 382 was tested as prototype for model J-1 with the following weights; wt. empty 1641; useful load 720; crew wt. 170; fuel & oil 210 lbs.; payload 340 lbs.; gross wt. 2361 lbs.; performance differences were as follows; landing speed 45; climb 750 ft. first min. at sea level; climb in 10 min. was 5900 ft.; ceiling 13,000 ft. This same airplane also served as the basis for the development of the first model J-2.

Construction details for the "Robin" model J-2 were typical to the other models, including the following. Rear bench-type seat was removable to provide cargo area of 32x47x41 inch dimensions; the baggage compartment

was an additional 12.5 cu. ft. with allowance for 50 lbs. Landing gear tread was 97 in., shock absorbing struts were oleo-spring, and wheels were 28x4. A metal propeller, wheel brakes, tail wheel, navigation lights, and hand-crank inertia-type engine starter were available as optional equipment. The next Curtiss development was the "Thrush" model J as discussed in the chapter for ATC # 236 in this volume.

Listed below are the only known entries of the "Robin" model J-2 as gleaned from registration records:

NC12H; Model J-2 (# 382) J 6-5-165
NC-790M;   ,,   ,,   (# 691)   ,,

Serial # 382 also listed as model J-3, but we have no information on this rare version; Robins with serial # 382 and upwards, eligible for modification to J-2 series.

Fig. 69. J-2 "Robin" differed in weights and fuel capacity from standard model J-1.

FOKKER "AMPHIBIAN", F-11A

Fig. 70. Fokker model F-11A amphibian; note novel sponsons.

Though one of the lesser known of the Fokker monoplanes, the F-11A "boat" was no doubt the most beautiful; a large and graceful craft of typical family resemblance. Riding at anchor, the F-11A was every inch a "ship" that gave off the feeling of distant shores and high adventure, comparable to the most trim of sailing vessels. Being an amphibious aircraft, the F-11A could operate equally well on land or on water, and therein lie its utility. As an air-ferry it could operate from water-front docks right in the heart of town, thus eliminating the usual ride to and from the airport. As a flying-yacht for the sportsman-pilot, the Fokker "Amphibian" was thoroughly at home on some landing strip close to hunting and fishing areas, or on the bay-waters of some play-land resort. This versatility of the "amphibian" was also ideal for men of business; Gar Wood, eminent power-boat builder and racing driver, commuted often in his craft between his boat-works and other business interests in the north, to his country home in the south. Decked out with comforts and extra conveniences, Gar Wood's F-11A was literally a "flying yacht". The prototype of the F11 series was built for H. S. Vanderbilt of New York and used strictly for pleasure jaunts in

Long Island waters. Of necessity, the F-11A carried a rather high price tag and this kept it from selling in any great number; the country was suffering from economic distress about this time and not very many people were able to buy expensive airplanes.

The Fokker F-11A was a high-wing cantilever monoplane of the "flying boat" type with seating for 8 in the forward portion of the enclosed all-metal hull; the engine, a 9 cyl. Wright "Cyclone" of 525 h.p. for this version, was tripod mounted in a streamlined nacelle, high above the wing in "pusher" fashion. A 3-bladed propeller was used to cut down the overall diameter and allow the nacelle to be placed closer to the wing. A retractable landing gear was incorporated into the streamlined sponsons jutting out from the hull; these "wing stubs" provided extra lift and also steadied the craft during water operations. Apparently these sponsons were not entirely satisfactory because they were soon discarded and small floats were placed out near the wing tips to keep the hull on even keel while in the water. In answer to demands for extra equipment without the penalty of lost performance, the gross weight was upped to 7200 lbs. and power increased to

575 h.p. For a much greater increase in all-round performance, "Tony" Fokker suggested the development of a powerful twin-engined "pusher-puller" version of the F-11A as shown; this he demonstrated often but no other examples of this high-performance "twin" were built. The type certificate number for the Fokker "Amphibian" model F-11A as powered with the 525 h.p. Wright "Cyclone" engine, was issued 9-5-29 and probably no more than 2 or 3 examples of this version were manufactured by the Fokker Aircraft Corp. Alfred A. Gassner, eminent aeronautical engineer of wide experience, was chief engineer for "Tony" Fokker at this time. A few years later, Gassner designed the famous "Baby Clipper" amphibian airplane for Fairchild Aircraft.

Listed below are specifications and performance data for the Fokker "Amphibian" model F-11A as powered with the 525 h.p. "Cyclone" engine; length overall 45'10"; hite wheels up 13'0"; hite wheels down 14'5"; wing span 59'0"; total wing area 550 sq.ft.; airfoil "Fokker"; wt. empty 4470; useful load 2430;

payload with 120 gal. fuel was 1449 lbs.; gross wt. 6900 lbs.; max. speed 112; cruising speed 95; landing speed 50; climb 800 ft. first min. at sea level; ceiling 11,500 ft.; gas cap. 120 gal.; oil cap. 12 gal.; cruising range at 28 gal. per hour was 400 miles; price at the factory was $42,000. for the amphibian, and lowered to $33,775. in June 1930; price for the flying boat less landing gear was $40,000. and lowered to $32,500. in June 1930. With extra equipment and plush interior arrangement, the F-11A weighed in at 4937 lbs. empty; useful load 2263; payload 1282; gross wt. 7200 lbs.; another version weighed in at 5065 lbs. empty; useful load 2135; payload 990 lbs. with allowance for 2 pilots; these heavier versions were the F-11AHB and mounted the Pratt & Whitney "Hornet" B of 575 h.p., and performance differences were as follows; max. speed 116; cruising speed 98; landing speed 55; climb 750 ft. first min. at sea level; climb in 10 min. 5800 ft.; ceiling 11,000 ft.; all other figures remained same.

The hull framework was built up of duralumin sections riveted together in truss

Fig. 71. Tony Fokker alighting from test flight in F-11A amphibian.

Fig. 72. Twin-engined F-11 for high performance; converted as personal plane for Tony Fokker.

Fig. 73. Fokker F11A also offered as flying boat; ideal as air-ferry.

form and covered with aluminum alloy sheet that was riveted to the framework; the vee-bottom hull with cabin dimensions of 59x60x 130 inches, had ample room for 8 seats or any number of other seating arrangements. The 4 forward seats were raised slightly above the level of the other four, with a 44 cu. ft. baggage compartment aft of the cabin section. The space aft of the main cabin could be converted to include a lavatory, a small galley, and sleeping quarters, which of course could not be used during flight. Entrance door to the main cabin was on the left side, with a narrow walk-way around the forward portion of the hull; all windows were of shatter-proof glass and dual controls were provided with a swing-over control wheel. The cantilever wing framework was built up of spruce and birch plywood box-type spar beams, with plywood wing ribs and stringers; the completed framework was covered in plywood veneer. The two fuel tanks were mounted in the wing, flanking each side of the hull; fuel was fed to the engine by an engine-driven fuel pump, with a hand-operated pump for emergency use. The long-leg landing gear of 173 inch tread used Rusco rubber-ring shock absorbing struts; the landing gear folded up out of the way for water operations. Wheels were 36x8 and Bendix brakes were standard equipment. The vertical fin of riveted duralumin framework was built integral with the hull and covered with aluminum alloy sheet; the

rudder, stabilizer, and elevators were of welded steel tube framework and covered in fabric. A 3-bladed ground adjustable metal propeller, navigation lights, electric inertia-type engine starter, anchor and ropes, life preservers, booster magneto, and 2 fire extinguishers, were standard equipment. Extra equipment and special custom interior arrangements were available on order. The next Fokker development was the parasol-winged model F14 as discussed in the chapter for ATC # 234 in this volume.

Listed below are Fokker F-11A entries that were gleaned from registration records:

NC-7887; Model F-11A (# 901) Cyclone.
NX-148H;   „   „   (# 902)   „
NC-127M;   „   „   (# 904) Hornet B.
NC-339N;   „   „   (# 906)   „

Serial # 901 as prototype, was arranged as 6 pl. craft with 450 h.p. P&W "Wasp" engine; serial # 901 later on Group 2 approval numbered 2-163 as 6 pl. F-11A Special with 525 h.p. Cyclone and 6000 lbs. gross; serial # 904 built for Gar Wood as 8 pl. F-11AHB with Hornet B of 575 h.p. and 7200 lbs. gross on Group 2 approval numbered 2-172; serial # 906 built as 10 pl. F-11AHB air-ferry with 575 h.p. Hornet B and 7200 lbs. gross on Group 2 approval numbered 2-200; the registration numbers for serial # 903 and # 905 are unknown.

KREUTZER "AIR COACH", K-5

Fig. 74. Kreutzer "Air Coach" model K-5 with 3 Kinner K5 engines.

The trim "Air Coach" model K-5 was the latest development in the Kreutzer light tri-motor design, an arrangement calculated to assure reliability of operation in an airplane especially designed for the smaller air-lines operating over rugged and desolate country, or the business man to whom would appeal the extra comfort and assurance of multi-engined flying. To sell the utility of air travel, manufacturers were deemed to stress safety and reliability of operation above all else; the assurance of reasonable safety and reliability would certainly be most appealing to those interested in air-travel, but who were yet largely concerned with the occasional chance of engine failure. The easiest way to insure the sense of safety and reliable operation in an airplane was the multi-engined configuration; multiple engines provided that sense of comfort and assurance to know that power was still available for continued flight, even in the case of failure to one engine. Large "twins" and "tri-motors" had been used extensively on air-lines for several years, now the trend was swinging towards the light multi-engined airplane. The Kreutzer "Air Coach", first introduced to the public in 1928, was among the first offered in this new concept. Offered in 2 different models during the earlier part of 1929 (refer to chapters for ATC # 170-171 in U.S. CIVIL AIRCRAFT, Vol. 2), the latest offering was exemplified in the model K-5 which was now more plush with somewhat higher performance.

The good-looking Kreutzer "Air Coach" model K-5 was a "baby tri-motor" high wing monoplane that was powered with 3 five cyl. Kinner K5 engines of 100 h.p. each. There was ample room and good comfort for six in a spacious and well appointed cabin, offering heat and ventilation, the safety of shatter-proof glass, with a promise for an extra margin of safety and performance provided by the three 100 h.p. engines. In comparison to previous models of the "Air Coach", the model K-5 now had a total of 300 h.p., which was a reserve of power that translated into higher performance, with ample power still

Fig. 75. Kreutzer "Air Coach" K-5 revived later as air transport T-6.

available for continued flight to a haven of refuge in the case of failure to any one engine. Well suited for feeder-line service over rough and desolate country, the K-5 "Air Coach" was used by lines in Arizona and Louisiana, which both abounded in treacherous terrain where an extra margin of safety, or even implied safety, would surely be appreciated by those on board. Business men found this new version very attractive and several were used in business promotion; the Kinner Engine Co. operated a Kreutzer K-5 in demonstration tests around the country. The type certificate number for the Kreutzer model K-5, as powered with three 100 h.p. Kinner K-5 engines was issued 9-6-29 and some 8 or more examples of this model were manufactured by the Joseph Kreutzer Corp. (Aircraft Division) of Los Angeles, Calif. In a slight re-organization of the company about this time, Joseph Kreutzer became chairman of the board; Howard Throckmorton was the president and treasurer; Albin K. Peterson was V.P. and chief of design and engineering.

The Kreutzer "Air Coach" series were designed by Albin K. Peterson who was well-noted in the California aircraft industry circles for some outstanding designs. In the early thirties, after the Kreutzer Co. suspended

production on the "Air Coach" series, Peterson designed the light "Meteor" sport monoplane which was an interesting craft, but failed to attract any substantial business because of the economical depression that still held sway in this country. In 1935, the Air Transport Mfg. Co. of Glendale, Calif. was organized with A. K. Peterson as V.P. and chief engineer, to revive the production of the Kreutzer K-5 (changed to model T-6) and the "Meteor" monoplane, but the new venture was not successful.

Listed below are specifications and performance data for the Kreutzer "Air Coach" model K-5 as powered with three 100 h.p. Kinner K5 engines; length overall 33'6"; hite overall 9'6"; wing span 48'10"; wing chord 84"; total wing area 315 sq.ft.; airfoil Goettingen 398; wt. empty 2745; useful load 1698; payload with 85 gal. fuel was 951 lbs.; gross wt. 4443 lbs.; max. speed 130; cruising speed 110; landing speed 45; climb 950 ft. first min. at sea level; ceiling 17,000 ft.; gas cap. 85 gal.; oil cap. 9 gal.; cruising range at 18 gal. per hour was 520 miles; price at the factory was approx. $18,500. The following figures are for later version as designated model T-6; wt. empty 2828; useful load 1672; payload with 85 gal. fuel was 925 lbs.; gross wt. 4500 lbs.; due to careful streamlining and the use of "wheel pants" (wheel fairings), the max. speed was raised to near 140 m.p.h.; other figures remained more or less the same.

The fuselage framework was built up of welded chrome-moly steel tubing in truss form, faired to an oval cross-section with duralumin formers and fairing strips, then fabric covered. The cabin area was spacious and well appointed with provisions for cabin heat and ventilation; the cabin walls were sound-proofed and insulated with "Seapak". The cabin was fitted with two large entrance doors, shatter-proof glass, and dual wheel controls were provided; the large baggage compartment was to the rear of the cabin section, with provisions for a small lavatory and locker. The semi-cantilever wing framework, in two halves, was built up of spruce and plywood box-type spar beams with spruce and plywood truss-type wing ribs; the leading edge was covered with plywood and the completed framework was covered with fabric. The fuel supply was carried in 2 gravity-feed tanks mounted in the root end of each wing half; an oil tank of 3 gal. capacity was mounted in each engine nacelle. The landing gear of 162 inch tread was of the outrigger type using "Aerol" shock absorbing struts; wheels were 32x6 and Bendix brakes were standard equipment. A full swivel tail-wheel was a great help in ground maneuvering. The fabric covered tail-group was built up of welded

chrome-moly steel tubing; the fin was ground adjustable and the horizontal stabilizer was adjustable in flight. Navigation lights, metal propellers, fire extinguishers, first-aid kit, storage battery, and engine starters were also standard equipment. The Air Transport model T-6 version was quite typical but offered 9.50x12 semi-balloon tires with wheel fairings, and electric Eclipse engine starters. The next approved development designed by A. K. Peterson, was the "Meteor" P-2 which will be discussed in the chapters for ATC # 482 and # 488.

Listed below are Kreutzer model K-5 entries that were gleaned from registration records:

NC-9354;	Model	K-5	(# 104)	3 Kinner K5.
NC-983H;	„	„	(# 107)	„
NC-982H;	„	„	(# 108)	„
NC-243M;	„	„	(# 110)	„
NC-244M;	„	„	(# 111)	„
;	„	„	(# 112)	„
NC-187W;	„	„	(# 113)	„
NC-995Y;	„	„	(# 114)	„

Serial # 104 was first a model K-3 with 3 LeBlond 90; registration number for serial # 107 unverified; registration number for serial # 112 unknown; serial # 113 operated in Central America; serial # 114 later modified by Air Transport Mfg. Co. to model T-6.

Fig. 76. New Standard D-25A in Coast Guard colors; 5 place D-25A had 7 cyl. Wright J6 engine.

The familiar "Standard" biplane of the first World War period was a large but rather graceful airplane, a pleasant airplane that was quite docile and obedient. The "New Standard" biplane of some 10 years later, was almost like a re-incarnation of this old favorite, just done up in more modern dress. The "Hisso" (Hispano - Suiza) powered model D-24 was the first of this series and a very unusual airplane for this particular time, reminding one very much of the old Pitcairn, Sikorsky, Woodson, and other such biplanes that were built back in the 1923-1925 era. Airplanes that carried four passengers and a pilot, all out in the open, may have looked outmoded by now, but it was actually a very sensible arrangement, one that worked out quite well for carrying passengers on short joy-rides around the vicinity of the air-field. The occupants certainly enjoyed the thrill and the feel of the open cockpit, and 4 paying passengers enabled the operator to show a fairly decent profit on a busy week-end. The D-25 was actually designed in anticipation of the revival of the "barnstorming era" of a few years back; joy-riding was still quite popular with the average airport visitor who brought the wife and the kids to look around.

The New Standard model D-25A in discussion here, was typical to the previous model D-25 in all respects, except for the powerplant installation which was now the 7 cyl. Wright J6 engine of 225 h.p. The D-25A was also a 5 place open cockpit biplane seating four passengers in one large open cockpit up forward, and the pilot was seated in an open cockpit further back. Largely concerned with a substantial paying load and good short-field performance, the "New Standard" in this series were particularly endowed with a generous amount of efficient wing area; the huge upper wing had a graceful sweep and the small lower wing approached the sesqui-wing arrangement. Though large and quite bulky, the "New Standard" was a graceful and amiable airplane that offered a surprisingly good performance in spite of its over-sized proportions.

The model D-25A was groomed as the improved offering for 1930 but it is surprising that the model D-25 with Wright "Whirlwind" J5 engine (refer to chapter for ATC # 108 in U.S. CIVIL AIRCRAFT, Vol. 2) continued on in popularity and the model D-25A was built only in small number. Of the few examples of the model D-25A that were out in service, one was a veteran in Alaskan service that was modified from a Wright J5 powered D-25; a modification of this nature would be a fairly easy chore because all changes made were ahead of the firewall. The type certificate number for the "New Standard" model D-25A as powered with the Wright J6 engine of 225 h.p., was issued 9-7-29 and some 3 or more examples of this model were manufactured by the New Standard Aircraft Corp. at Paterson, New Jersey.

Listed below are specifications and performance data for the model D-25A as powered with the 225 h.p. Wright J6 engine; length overall 26'10"; hite overall 10'2; wing span upper 45'0"; wing span lower 32'6"; wing chord upper 70"; wing chord lower 50" wing area upper 240 sq.ft.; wing area lower 100 sq.ft.; total wing area 350 sq.ft.; airfoil Goetingen 533; wt. empty 2055; useful load 1290; payload with 62 gal. fuel was 712 lbs.; gross wt. 3345 lbs.; max. speed 110; cruising speed 96; landing speed 37; climb 760 ft. first min. at sea level; ceiling 18,000 ft.; gas cap. 62 gal.; oil cap. 5 gal.; cruising range at 12.7 gal. per hour was 440 miles; price at factory was first announced as $9700. but reduced to $7990. in July of 1930.

Of durable and rugged construction, the huge wing panels of the "New Standard" were of the normal wood spar beam and wood wing rib make-up, but the fuselage was a very novel framework. The fuselage was built up of duralumin angles and open section members that were riveted and bolted together into a simple and strong framework. Quite a new concept at this time, this was a simple construction method that could be easily repaired in the field without a large layout of special tools and special equipment, easily repaired with tools one might find in any good tool box. The split-axle landing gear of 96 inch tread used oleo-spring shock absorber struts; wheels were 30x5 and Bendix brakes were available. A wooden propeller was standard equipment, but a metal propeller and inertia-type engine starter were optional. A coupe-top was reported developed for this model to convert it to a cabin-type airplane (model D-25C), but little else is known of this conversion. The next development in the "New Standard" series was the 3 place model D-26A; see chapter for ATC # 225 in this volume.

Listed below are model D-25A entries that were gleaned from registration records:
NC-9190; D-25A (# 135) J6-7-225
NC-930V;   „   (# 152)   „
NC-33K;   „   (# 201)   „
NC-38K;   „   (# 204)   „
NC-150M;   „   (# 205)   „

Serial # 135 operating with Alaskan Airways, first had Wright J5 as model D-25; serial # 152 later modified with Wright J6 of 300 h.p. as model D-25B.

Fig. 77. D-26A was typical to craft shown (D-26) but had 7 cyl. Wright J6 engine.

The very distinctive "New Standard" series were conceived and designed by Charles Healy Day, who will be remembered as one of our leading aeronautical engineers from this period. Charles Day had designed and developed a good number of airplanes, many of them outstanding, but probably his most famous design was the beloved old "Standard" biplane of World War 1 vintage. This swept-wing biplane with its myriad of wing struts and bracing wires, was used by barn-storming pilots the country over, used by many flying-schools and even early air-lines for nearly 10 years afterward. True, the old "Standard" biplane had many short-comings but it did lend itself to all manner of modification, which often made it into a very useful machine. Very typical, in versatility at least, the "New Standard" series was also a basic design that could be easily arranged into various versions to suit a purpose; the model D-26A was a good example of this statement.

The model D-26 was but a modification of the basic D-25 design, that is, the D-26 was typical in all respects except that it was a 3 place airplane with room left over for an ample amount of baggage and yet a sizeable load of cargo, in addition to the passengers carried. Two passengers were seated in the front open cockpit and the pilot was seated in the rear open cockpit; a hatch-covered compartment in the forward portion of the fuselage had capacity to carry either a large amount of personal baggage, beyond the normal 30 lb. limit, for extensive traveling, or a cargo payload of nearly 450 lbs. A smaller baggage and tool compartment of 4 cu. ft. capacity was located in the turtle-back section of the fuselage behind the pilot's cockpit. The model D-26A was typical to the earlier D-26, except for its powerplant installation which was now the 7 cyl. Wright J6 engine of 225 h.p. The performance of this 3 place version was typical except for a slightly higher maximum and cruising speed, which was largely due to the lesser amount of turbulence created by the smaller front cockpit. The 3 place version in both D-26 and D-26A models was no doubt designed for shuttle-type service on the smaller air-lines, or perhaps for extensive traveling in the line of business promotion, but a great need for this particular type of airplane never did materialize. The type certificate number for the model D-26A as powered with the Wright J6 of 225 h.p. was issued 9-7-29 and probably no more than one example of this model was built by the New Standard Aircraft Corp. of Paterson, N. J.

Listed below are specifications and performance data for the "New Standard" model D-26A as powered with the 225 h.p. Wright J6 engine; length overall 26'10"; hite overall 10'2"; wing span upper 45'0"; wing span lower 32'6"; wing chord upper 70"; wing chord lower 50"; wing area upper 240 sq.ft.; wing area lower 110 sq.ft.; total wing area 350 sq. ft.; airfoil Goettingen 533; wheel tread 96"; wheels 30x5; wt. empty 2055; useful load 1345; payload of 2 passengers and 438 lbs. cargo, or total payload of 768 lbs with 62 gal. fuel; gross wt. 3400 lbs.; max. speed 114; cruising speed 100; landing speed 37; climb 750 ft. first min. at sea level; ceiling 18,000 ft.; gas cap. 62 gal.; oil cap. 5 gal.; cruising range at 12.7 gal. per hour was 450 miles; price at the factory with standard equipment was approx. $9000.

Except for cockpit arrangement and baggage-cargo compartment, the construction details of the model D-26A were typical as for all other models in the "New Standard" series; for further details refer to previous chapters discussing the various models. The interesting arrangement of the "New Standard" biplane with its large upper wing and smaller lower wing, was approaching the "sesqui-plane" configuration; due to the smaller area of the lower wing and its great amount of positive stagger in relation to the upper wing, the pilot was able to see his wheels easily from his position in the rear cockpit. This was helpful to the pilot on take-off and landings; in fact, visibility from the "New Standard" was excellent in all directions. In view of the fact that these biplanes had so much wing area, one is reminded that recovering these wings was a chore requiring nearly 100 yards of cotton fabric, not to mention the hundreds of yards of pinked-edge tape, the gallons and gallons of dope, and so on. The next development in this series was the single-place model D-27A, which is discussed in the chapter for ATC # 226 in this volume.

Listed below is the only known entry of the model D-26A as gleaned from registration records:

NC-35K; D-26A (# 202) J6-7-225.

Serial # 202 later modified to model D-26B with 300 h.p. Wright J6 engine for Electro Sky Ads of New York.

Fig. 78. Lineage of "Standard" J-1 shown, reflects back through "new Standard" of some 10 years later.

Fig. 79. D-27A was typical to craft shown but had 7 cyl. Wright J6 engine.

Yet another version of the basic "New Standard" biplane series was the model D-27A, which was typical to the models D-25A and D-26A in most all respects, except that it was a single-place airplane, having an open cockpit for the pilot and large hatch-covered compartments of 56 cu. ft. capacity up in the forward section of the fuselage; compartments for carrying air-mail and air-express cargo. The large and convenient cargo hold was spacious enough even for bulky loads, and could carry a payload up to 800 lbs. Performance of the model D-27A should have been slightly better in the top speed and cruising speed range because of the absence of large and gaping open cockpits, as on the D-25A, which created turbulence and drag of considerable proportions. Due to the better streamlining of the forward fuselage section, the model D-27A would cruise and top-off at least 5 m.p.h. faster. Other inherent characteristics such as quick take-off and slow landing speeds, which added up to good short field performance, were typical.

The model D-27A was more or less identical to the previous model D-27 in all respects except for its powerplant installation which was now the 7 cyl. Wright J6 engine of 225 h.p. Four of the Wright J5 powered model D-27 were built nearly a year earlier for Clifford Ball, but there seems to be no evidence of the number built in the model D-27A version. Actually, any existing model D-27 could have easily been modified to the D-27A version by merely changing engines; engine mounts and engine cowling would have to be arranged to suit the installation. No particular advantage would have been gained by this conversion, except perhaps that the Wright J6 series engine was improved considerably over the earlier Wright J5, and therefore offer more reliability and less care in extended service. The type certificate number for the model D-27A as powered with the 225 h.p. (R-760) Wright J6 engine, was issued 9-7-29; registration records failed to reveal any entries of this particular version. For a comparison of all the various models in the "New

Standard" series, we suggest reference to the chapters for ATC # 107 - 108 - 109 - 110 in U.S. CIVIL AIRCRAFT, Vol. 2, plus discussions in this volume contained in chapters for ATC # 224 - 225 - 226.

Listed below are specifications and performance data for the model D-27A as powered with the 225 h.p. Wright J6 engine; length overall 26'10"; hite overall 10'2"; wing span upper 45'0"; wing span lower 32'6"; wing chord upper 70"; wing chord lower 50"; wing area upper 240 sq.ft.; wing area lower 110 sq.ft.; total wing area 350 sq.ft.; airfoil Goettingen 533; wt. empty 2055; useful load 1345; payload with 62 gal. fuel was 768 lbs.; gross wt. 3400 lbs.; these weights would vary somewhat with the addition of night-flying equipment; max. speed 118; cruising speed 105; landing speed 37; climb 750 ft. first min. at sea level; ceiling 18,000 ft.; gas cap. 62 gal.; oil cap. 5 gal.; cruising range at 12.7 gal. per hour was 470 miles; basic price at the factory was $9000., plus cost for extra or special equipment.

Except for cockpit arrangements, construction details were typical for all models in the "New Standard" series discussed here. The mail-carrying version was completely wired for lights, with two large landing-lights built into the leading edge of the upper wings for night-flying operations. Other features that are typical of all models in this series are, fuel tanks in the center-section panel of the upper wing, a small baggage and tool compartment in the turtle-back section of the fuselage, and a large exhaust collector ring with extended tail pipe that expelled fumes well below the cockpit level. The large and graceful wings on the "New Standard" were semi-elliptical in plan-form and had an upward sweep at the tips, which acted as extra dihedral angle. Dihedral on the upper wing was 2 deg., dihedral on the lower wing was 4 deg., gap between panels was 72 in., and stagger between panels was 34 in.; these geometrics added up to a stable, easy to fly, and well-behaved airplane.

In the chapter for ATC # 110 of U.S. CIVIL AIRCRAFT, Vol. 2, we stated that it could not be determined whether Clifford Ball had put his four D-27 into active service; information received since has proven that the D-27 were in active service on the mail-route from Cleveland-Youngstown-Pittsburgh for a time at least. The D-27 was used mostly on the night mail-run but the pilots actually preferred the comforts of the closed cabin airplanes; 3 of the D-27 were lost in accidents at different times, and the 4th was finally retired from mail service. The simple and rugged structure of the "New Standard" was certainly conducive to longevity, so it is not surprising then that several of these fine old airplanes lasted for a good many years; a few are in service yet. The next development in the "New Standard" series was the model D-29S sport-trainer which will be discussed in the chapter for ATC # 324.

Fig. 80. Head-on view of D-27A showing 7 cyl. Wright J6 engine and generous wing area.

A.T.C. #227
(9-9-29)
LOCKHEED "VEGA", 5-B

Fig. 81. Lockheed Vega 5-B with 450 H.P. Wasp engine.

The fabulous Lockheed "Vega" with its flawless lines was still a magic name in aviation circles and a combination even yet hard to beat, despite the experience and knowledge that had been gained in aircraft design these past few years. Much of the history of speed and distance in the air had already been written by the "Vega" and much was yet to be written in the next few years to come. Bolstered by its many accomplishments, the cigar-shaped Lockheed "Vega" was a glamorous airplane and by now, several were as famed as their illustrious pilots, out-classing just about anything that flew. Added now to this impressive line-up of various "Vega" versions already in production, was the 7 place model 5-B, a load-toting combination of pure symmetry that was probably the most famous "Vega" model in the whole star-studded line-up. Many famous "Vega" record-breakers were basically of the model 5-B configuration and modified only to the extent necessary for a particular purpose; the "Winnie Mae" that twice shattered the round-the-world record was a model 5-B to begin with and it was no doubt the most famous and the greatest of them all. Other well known 5-B versions that earned varying degrees of

fame and fortune were the Stanavo "Flying Trade-Marks", the "Century of Progress", the "City of New York", "Akita", the "Cherokee" Amelia Earhart's "Little Red Bus", and several others. Arranged as a 7 place craft, the 10-windowed model 5-B was ideal for passenger service requiring medium capacity and was used on a good number of air-lines for fast schedules; these lines were the fastest in the world. Among the many lines that used the model 5-B in regular scheduled service were Alaska-Washington Airways, Bowen Air Lines, Braniff, Hanford Tri-State, Midland Express, Wedell-Williams, and also C.A.T. and L.A.M.S.A. in Mexico. It would be hard to decide, because all were winners in one way or another, but if a selection just had to be made on merit, it is quite likely that the model 5-B would prove itself as the greatest "Vega" version of them all.

The Lockheed "Vega" model 5-B as pictured here, was a pure cantilever high winged monoplane seating 7; although primarily designed as an executive-transport and small air-liner for fast transportation on the shorter routes, it was hard to break the habit befallen to all "Lockheeds". It too was destined to be used for all sorts of speed, altitude, and distance records. It was only some years later, when the "Vega" finally became out-classed by ships of new design, that it was allowed to render the service it was planned for. Powered with the thundering 9 cyl. Pratt Whitney "Wasp" engine of 450 h.p., the model 5-B had the advantage of extra seating, while performance and general behavior remained typical or even somewhat better than previous models of the "Vega". Equipped with the bulbuous N.A.C.A. type low-drag engine cowling and long slender "tear drop" wheel pants, the "Vega" became almost a pure streamlined shape that translated into effortless speed and extra range far beyond the average. Because of the extra seating, 5 windows were arranged down each side of the cabin section and this identifiable mark set the model 5-B somewhat apart from other "Vega" versions which only had 4 in a row; that is, until the introduction of the model 5-C which also had five windows but it was set apart by a liberal increase in the fin and rudder area. As a shake-down trip for the new "Vega", the model 5-B was flown in the National Air Tour for 1929 by assistant test-pilot Wiley Post who finished in 17th place but was easily the sensation at stops along the far-flung tour. The type certificate number for the 7 place "Vega" model 5-B as powered with the 450 h.p. "Wasp" engine, was first issued on 9-9-29 for landplane versions and amended 1-23-30 for the seaplane version on Edo K floats. Some 30 or more examples of this model were manufactured by the Lockheed Aircraft Corp. at Burbank, Calif.; a division of the Detroit Aircraft Corp. of Detroit, Mich.

Listed below are specifications and performance data for the Lockheed "Vega" model 5-B as powered with the 450 h.p. "Wasp" engine; length overall 27'6"; hite overall 8'6"; wing span 41'0"; wing chord at root 102"; wing chord at tip 63"; total wing area 275 sq.ft.; airfoil at root Clark Y-18; airfoil at tip Clark Y-9.5; wt. empty 2490 lbs.; useful load 1775; payload with 70 gal. fuel was 1110 lbs.; payload with 100 gal. fuel was 930 lbs; gross wt. 4265 lbs.; max. speed (with N.A.C.A. cowl) 180; cruising speed 155; landing speed 55; climb 1275 ft. first min. at sea level; ceiling 20,000 ft.; gas cap. max. 100 gal.; oil cap. 8-10 gal.; cruising range at 22 gal. per hour was 690 miles at max. gas cap.; the following figures are for the 5-B as a seaplane mounted on Edo K twin-float gear; wt. empty 2820; useful load 1930; payload with 100 gal. fuel was 1085 lbs.; gross wt. 4750 lbs.; max. speed 170; cruising speed 145; landing speed 65; climb 1050 ft. first min. at sea level; ceiling 17,500 ft.; gas cap. 100 gal.; oil cap. 10 gal.; cruising range at 22 gal. per hour was approx. 600 miles; price at the factory field was $18,895. for the landplane; seaplane was about $1500. extra.

The fuselage framework of the 5-B was a wooden monococque construction built up of two "plywood shells" that were assembled over spruce annular rings. The inner walls of

Fig. 82. Vega 5-B arranged with seating for 7.

Fig. 83. Vega 5-B on floats; note extra windows.

the cabin were lined in a heavily-padded pig-skin which gave the interior an elegant sense of richness; there were four single adjustable seats and one double seat in the rear. A baggage compartment of 12 cu. ft. capacity was to the rear of the cabin section with allowance for 120 lbs. for the landplane and 95 lbs. for the seaplane. The cantilever wing framework was built up of spruce and ply-wood box-type spar beams with spruce and plywood web-type wing ribs; the completed framework was covered with 3/32 in. ply-wood veneer. The long-leg landing gear used "oleo" shock absorbing struts and was quickly detachable for installation of float gear; wheels were 32x6 and Bendix brakes were

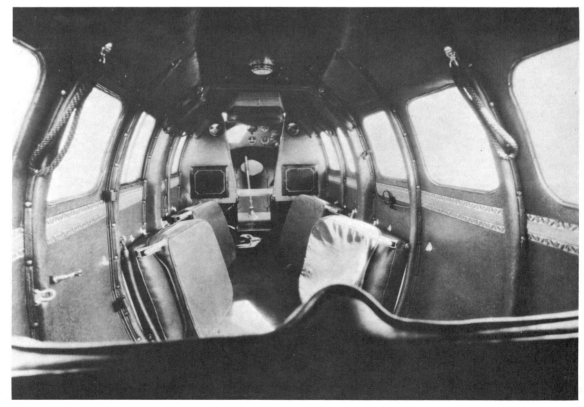

Fig. 84. Interior of 7 place Vega; upholstery was padded pig-skin.

standard equipment. The cantilever tail-group, now slightly larger in area, was all-wood and of similiar construction to the wing; the horizontal stabilizer was adjustable in flight. A metal propeller, electric inertia-type engine starter, navigation, cockpit, and cabin lights, cabin and cockpit heaters, fire extinguisher, and first-aid kit were provided as standard equipment. The next "Vega" development was the 7 place model 2-A as described in the chapter for ATC # 252 in this volume.

Listed below are Vega 5-B entries that were gleaned from various registration records:

Fig. 85. N.A.C.A. cowling and wheel pants boosted Vega speed to easy 180 M.P.H.

NC-2845;	Vega 5-B	(# 61)	Wasp 450.
NC-2846;	„	(# 62)	„
NC-858E;	„	(# 66)	„
NC-868E;	„	(# 68)	„
NC-892E;	„	(# 74)	„
NC-336H;	„	(# 81)	„
NC-504K;	„	(# 90)	„
NC-48M;	„	(# 100)	„
NC-49M;	„	(# 101)	„
NC-534M;	„	(# 103)	„
NC-536M;	„	(# 105)	„
NC-537M;	„	(# 106)	„
NC-540M;	„	(# 109)	„
NR-500V;	„	(# 112)	„
NC-105N;	„	(# 117)	„
NC-106N;	„	(# 118)	„
NC-102W;	„	(# 119)	„
NC-103W;	„	(# 120)	„
NC-104W;	„	(# 121)	„
NC-105W;	„	(# 122)	„
NC-106W;	„	(# 123)	„
NC-107W;	„	(# 124)	„
NC-152W;	„	(# 125)	„
NC-160W;	„	(# 126)	„
NC-161W;	„	(# 127)	„
NC-162W;	„	(# 128)	„
NC-176W;	„	(# 129)	„
NC-905Y;	„	(# 133)	„
NC-926Y;	„	(# 134)	„
NC-997N;	„	(# 139)	„

Serial # 22 - 54 - 60 later converted to 5-B; serial # 101 converted to 5-B. Special on Group 2 approval 2-274; serial # 119-120 were seaplanes; serial # 61 - 62 later as XA-BHJ and XA-BHA in Mexico; serial # 90 later as XA-BHC; serial # 97 sometimes listed as 5-B (NC-46M); serial # 100 later as XA-BHK; serial # 103 later as XA-BHI and # 125 as XA-DAH; nearly all ships later converted to model 5-C configuration.

GREAT LAKES "SPORT", 2-T-1A

Fig. 86. Great Lakes 2-T-1A on test, Charlie Meyers flying.

The cocky little "Great Lakes" swept-wing biplane was probably one of the most distinctive-looking and most easily recognizable airplanes of this period; a heart-warming combination of dash and performance that made an association with this craft a memorable experience and offered many private-owners the chance to play at sportsman-pilot on a very modest budget. A little bit of genius, a stroke of good luck, put them all together and we have an airplane that has had a long span of exciting history and a country-wide popularity that hasn't waned a bit, even to this day. Contrary to the general belief in later years, the "Great Lakes" sport-trainer as first developed was a normal straight-wing biplane and did not have the characteristic sweep-back in the upper wing; this addition of sweepback was first a feature of necessity that later became such a distinguishing trade-mark. (For detailed explanation see chapter for ATC # 167 of U.S. CIVIL AIRCRAFT, Vol. 2). Formally introduced to the public at the Detroit Air Show of March 1929, the magnetic personality and the sprightly performance of the Great Lakes "Sport" was soon on everyone's lips and the good news spread throughout flying circles like wild-fire. Production lines at first could not cope with the

demand because the "Great Lakes" was bought up as fast as they could be built and the company soon had a back-log of some 200 orders.

Cocky and determined, the "Great Lakes" sport-trainer could not be considered gentle but it had no great vices and a good steady hand could transform its playful nature into an experience of lasting pleasure. As a case in point, we shall never forget the remark of one pilot — "I feel like I've just fallen in love"! Because of its nature the "Sport" could not tolerate timidity but it was easy to understand and it radiated with eagerness; a proper understanding between a pilot and this airplane usually transformed into mutual admiration and a lasting friendship. That the "Great Lakes" was extremely capable and only too willing to do as bid, can be reflected back from the many outstanding records and achievements that were left behind during its active life. In Aug. of 1929 "Tex" Rankin flew a 2-T-1 on a "three flags" non-stop flight from Canada through the U.S. and into Mexico averaging more than 18 miles per gallon at 103 m.p.h. on the 1350 mile flight. In Jan. of 1930 using the same airplane, "Tex" established a world record of 19 "outside loops"; an abnormal maneuver that really tested the mettle of an aircraft and stamina of the pilot. Later that year several other records were either set or broken by the "Great Lakes", the following year saw many more achievements, and so did the next and the next. A full record of achievements by this airplane for a span of some twenty years would be material for a book in itself. The capabilities of this airplane were only limited by the power available so many pilots in later years modified the sport-trainer with engines of almost double the original horsepower; the performance unleashed in these combinations was rather hard to beat. The robust framework, the hardy nature, and the popularity of the "Great Lakes" biplane had a tendency to promote its longevity so it is not surprising that nearly 100 were still flying some 10 years later and a good number are flying even yet, with many more destined to be rebuilt if they can only be found.

The first production version of the "Great Lakes" sport-trainer was the model 2-T-1 (ATC # 167) as powered with the "Cirrus" Mk 3 engine of 85 h.p.; a slightly improved version as approved under this certificate (ATC # 228) was powered with the American Cirrus engine of 90 h.p. and designated the 2-T-1A. The 2-T-1A was also a 2 place open cockpit biplane of diminutive proportion that seated two in tandem and was more or less typical to the 2-T-1 in most all respects except for the change in powerplants, a slight refine-

ment in the landing gear arrangement and a beefing-up of the tail structure. For 1931 the 2-T-1A was modified again with several changes that altered the face of things; the engine cowling was now deepened and rounded off into a buxom but somewhat simpler installation, the area of the fin and rudder was increased to promote more positive control at lower speeds and the hi-pressure tires were often replaced with donut-type low pressure air-wheels. The performance and general behavior remained more or less the same throughout the series but many have voiced a slight preference for the earlier type. The American Cirrus engine manufactured in this country by an affiliated organization of the G.L.A.C. was a slight refinement of the British "Cirrus" Mk. 3 engine and its power output was raised by some 5 h.p. although the increase was hardly noticeable. Of a 4 cyl. in-line arrangement, the air-cooled "Cirrus" was easy to streamline and well adapted for use in the small trainer-type airplane; its most memorable characteristic was perhaps its sound, a sound that was a rhythmic purr of pleasing tone and of considerable volume, especially with "short stacks".

The Great Lakes Aircraft Corp. which had been formed late in 1928 introduced the 2-T-1 sport-trainer in the early part of 1929 and soon had a back-log of some 200 orders. By mid-year of 1929 some 750 employees were busily engaged in turning out airplanes and the back-log of orders had zoomed to 700. The

Fig. 87. Experimental version with skis, winter enclosure and "wing slots."

Fig. 88. Nimble "Sport" also popular with flying salesmen.

80 acre plant-site with an adjoining air-field was centered about a plant area of some 90,000 sq.ft.; production lines and assembly lines were arranged for a continuous flow of airplanes, testing laboratories were busy analyzing new materials and testing structures, the engineering department was busy with air-planes for the future, and the air-field was humming with airplanes leaving on delivery flights. It seemed like a rosy future for the G.L.A.C. but the stock-market crash of late 1929 created an economic depression that dealt the company a severe blow; literally hundreds of orders were cancelled and humming production lines almost came to a halt. Reeling from the mortal blow but de-termined to carry on, the G.L.A.C. bided time. A drastic cut in prices was intended to stimulate buying but the expected flood of orders was only a trickle; from then on it was a struggle for existence. The type certificate number for the "Great Lakes" model 2-T-1A as powered with the American Cirrus engine was issued 9-17-29 and upwards of 200 examples of this model were manufactured by the Great Lakes Aircraft Corp. at Cleveland, Ohio. Col. Benjamin F. Castle was the presi-dent; Chas. F. VanSicklen was sales manager; Capt. Holden C. Richardson was chief of engineering; P. B. Rogers and Earl Stewart were project engineers; and Chas. W. Meyers was preliminary-design engineer, sales engi-neer and chief pilot in charge of test and development. In a slight reorganization of top-

level officials in 1931, Chas. F. Barndt became president; P. B. Rogers became chief of engineering; and Edw. Rembert became sales manager.

Listed below are specifications and per-formance data for the "Great Lakes" model 2-T-1A as powered with the 90 h.p. Ameri-can Cirrus engine; length overall 20'4"; hite overall 7'11"; wing span upper and lower 26'8"; wing chord both 46"; wing area upper 97 sq.ft.; wing area lower 90.6 sq.ft.; total wing area 187.6 sq.ft.; airfoil M-12; wt. empty 1102; useful load 578; payload with 26 gal. fuel was 243 lbs.; gross wt. 1580 lbs.; max. speed 115; cruising speed 97; landing speed 45; climb 780 ft. first min. at sea level; climb to 5000 ft. was 8 min.; ceiling 12,500 ft.; gas cap. 26 gal.; oil cap. 2 gal.; cruising range at 6.3 gal. per hour was 375 miles; price at the factory field was $4990., lowered to $3985., reduced to $3150., and then drastically cut to $2985. in 1931. The 1931 model of the 2-T-1A was typi-cal to figures above except as follows; max. speed 110; cruising speed 94; stall speed 50.

The construction details and general ar-rangement for the Great Lakes model 2-T-1A were typical to the model 2-T-1 as described in the chapter for ATC # 167 of U. S. CIVIL AIRCRAFT, Vol. 2 with addition of the following. The cockpit interiors were bare in the trainer and upholstered in the deluxe sport models; the baggage compartment had

Fig. 89. Test hop of 2-T-1A on floats over Cleveland, O.

allowance for 13 to 27 lbs. Ailerons and elevators were operated by steel push-pull tubes; the larger fin and rudder on later models increased overall length of fuselage by 6 inches. Landing gear of 70 inch tread used "Aerol" (air-oil) shock absorbing struts; wheels were normally 24x4 and Bendix brakes were available. Size 22x6, 22x10, or 26x8 low-pressure airwheels were also available. Tailskid was of the spring-leaf type with removable manganese "shoe"; a 6 x 2 tail-wheel was also available. The fabric covered tail-group was built up of duralumin tubular spars with stamped-out dural former ribs; the vertical fin was fixed and the horizontal stabilizer was adjustable in flight. The next develop-

Fig. 90. Beautiful 2-T-2 "Speedster" was racing version of 2-T-1A "Sport."

ment in the "Great Lakes" sport-trainer series was the model 2-T-1E with inverted Cirrus engine as described in the chapter for ATC # 354.

Listed below is a partial listing of Great Lakes model 2-T-1A entries as gleaned from registration records:

NC-433Y;	Model 2-T-1A	(# 150)	Amer.Cirrus
NC-434Y;	„	(# 151)	„
NC-435Y;	„	(# 152)	„
NC-436Y;	„	(# 153)	„
NC-437Y;	„	(# 154)	„
NC-807Y;	„	(# 176)	„
NC-808Y;	„	(# 177)	„
NC-809Y;	„	(# 178)	„
NC-810Y;	„	(# 179)	„
NC-811Y;	„	(# 180)	„
NC-11128;	„	(# 184)	„
NC-11129;	„	(# 185)	„
NC-11131;	„	(# 186)	„
NC-11139;	„	(# 187)	„
NC-11149;	„	(# 188)	„
NC-302Y;	„	(# 189)	„
NC-303Y;	„	(# 190)	„
NC-305Y;	„	(# 192)	„
NC-307Y;	„	(# 194)	„
NC-308Y;	„	(# 195)	„
NC-309Y;	„	(# 196)	„
NC-310Y;	„	(# 197)	„
NC-311Y;	„	(# 198)	„
NC-312Y;	„	(# 199)	„
NC-313Y;	„	(# 200)	„

This approval eligible for serial # 124 - 184 - 185 - 186 - 187 - 189 and up; all serial numbers eligible under ATC # 167 (# 7 to # 264) also eligible under this approval when changed to conform; standard allowable gross weight raised to 1618 lbs.

Fig. 91. 1931 model of 2-T-1A "Sport," various changes altered personality to some extent.

Fig. 92. General "Cadet" model 111-C with 110 H.P. Warner engine; arrangement was
ideal for training or sport.

The General Airplanes "Cadet" trainer basically resembled the distinctive "Aristocrat" design, except that it was arranged as a practical 2 place open cockpit parasol monoplane; a trim craft of very pleasant lines, as evident in the illustration, and apparently well suited for use as a sport-craft or a trainer. Not otherwise unusual, the "Cadet" was of clean simple lines and rugged structure, offered very good visibility and the aerodynamic arrangement suggested good behavior. The model 111-C as discussed here was powered with the 7 cyl. Warner "Scarab" engine of 110 h.p.; this engine was selected as the standard powerplant for this series, but the 4 cyl. Wright-Gipsy of 90 h.p. and the 7 cyl. Yankee-Siemens of 113 h.p. were also offered as optional installations. The General "Cadet" was said to have been a well-behaved airplane in all respects, but its rarity leaves much of its personality and operating history in question. Of the two examples that were built, one was serving on the flight line of the Silver Fox Air Lines flying-school as a basic trainer.

In the latter part of 1930, feeling the progressive pinch of the "depression", the General Airplanes Corp. was forced to vacate its plant in Buffalo and transfer all activities to a leased hangar on Roosevelt Field in Mineola, L. I., New York. Here, small-scale production was resumed on an improved version of the "Aristocrat" cabin monoplane and development was begun on the all-cargo (model 107) "Mailplane". In a plan to lower prices and thereby stimulate airplane buying, G.A.C. eliminated all of their distributors and dealers in a from-factory-to-buyer program; buyers were to be refunded the normal sales commission, which in a sense, amounted to price cuts of up to 30 per cent. Needless to say, this last-ditch program actually had very little effect in producing volume sales; not very many people in this country could afford to buy airplanes at this particular time.

In general description, the "Cadet" model 111-C sport-trainer was a 2 place open cockpit monoplane seating two in tandem; its "parasol wing" was a practical and sensible arrangement that offered excellent visibility, and inherently stable flight characteristics. Powered with a 7 cyl. Warner engine of 110 h.p., its power reserve should have been ample to guarantee a lively personality, as the performance figures given will bear out. Had this airplane been introduced under different circumstances at another time, one somehow gets the feeling that it would have been well accepted and extremely popular, both as a trainer and a sport-type craft for the private owner. The "Cadet" model 111-C received its type certificate number on 9-17-29 and only 2 examples of this model were manufactured by the General Airplanes Corp. at Buffalo, New York. In the transfer to quarters at Roosevelt Field, many of the company officials were shuffled around and replaced; A. Francis Arcier, who had been V.P. in charge of engineering, left soon after to join the Waco Aircraft Co. as their chief engineer.

Listed below are specifications and performance data for the General Airplanes "Cadet" model 111-C as powered with the 110 h.p. Warner "Scarab" engine; length overall 25'5"; hite overall 8'4"; wing span 36'4"; wing chord 72"; total wing area 202 sq.ft.; airfoil GAC-500; wt. empty 1206; useful load 535; payload with 22 gal. fuel was 215 lbs.; 170 lbs. was allowed for each occupant and 40 lbs. of baggage; gross wt. 1741 lbs.; max. speed 120; cruising speed 100; landing speed 40; climb 800 ft. first min. at sea level; climb in 10 min. 6800 ft.; ceiling 14,000 ft.; gas cap. 22 gal.; oil cap. 3 gal.; cruising range at 6 gal. per hour was 325 miles; price at the factory was first $5000., later reduced to $4000., and then to $3550. Prototype of the "Cadet" before certification, weighed in as follows; wt. empty 1126; useful load 511; gross wt. 1650; performance and payload were affected by the difference.

The fuselage framework was built up of welded chrome-moly steel tubing in a rigid truss form, lightly faired to shape and fabric covered. Individual cockpits were faired with a removable aluminum panel and a large door with convenient step was provided for ease of entry to the front cockpit; a baggage compartment behind the rear seat, was allowed a maximum of 40 lbs. Dual joy-stick controls were provided and seats had deep wells for parachute pack. The semi-cantilever wing framework, almost identical to that of the "Aristocrat", was built up of laminated spruce spar beams of heavy section and spruce and plywood truss-type wing ribs; the leading edge was covered with dural sheet and the completed framework was covered with fabric. The ailerons were built up of chrome-moly steel tube spars and duralumin sheet ribs, covered with fabric; for more positive control, the ailerons were of the offset-hinge type to provide quick response and aerodynamic balance. The wing was perched atop 4 inverted vee struts to form the center-section cabane; the outer wing braces were of heavy gauge chrome-moly steel tubing in a streamlined form. For the maximum in visibility, there was a large cut-out in the trailing edge of the wing just over the rear cockpit; a large sky-light in the center-section panel allowed visibility upward from the front cockpit. One fuel tank was mounted in the root end of the right wing half; oil tank was mounted in the engine compartment. The fabric covered tail-group was built up of welded chrome-moly steel tubing; the fin was ground adjustable and the horizontal stabilizer was adjustable in flight. For less stick and rudder pressure and quicker response, both rudder and elevators were fitted with aerodynamic "balance horns". The split-axle landing gear of 84 inch tread was built up of heat-treated aluminum alloy cantilever legs that were faired to a streamlined shape by sheet metal cuffs; the

Fig. 93. General "Observer" was twin-engined photo plane.

Fig. 94. Parasol wing on "Cadet" offered good visibility.

shock absorbers were rubber "donut rings" in compression. Main wheels were 28x4 or 30x5; individual wheel brakes and a steerable tail-wheel provided excellent ground maneuvering. The engine was well muffled by a large-volume collector-ring which kept the noise level at a minimum. Bendix wheels and brakes were standard equipment; navigation lights, an inertia-type engine starter, and a metal propeller were available as optional equipment. In mid-1930 plans were laid to offer several variations on the basic "Cadet" design; the "Cadet Trainer" with 165 h.p. Continental engine was offered for $3990.; a single-seated version of the "Cadet" with 165 h.p. Continental engine was called the "Postman" with a basic price of $4500.; a deluxe sport-type version of the "Cadet" with 165 h.p. Continental engine was called the "Sportsman" with a price tag of $5000. The next development in the "General Mono-planes" series was the "Aristocrat" model 102-F as powered with the 165 h.p. Continental A-70 engine; this version was built under a Group 2 approval numbered 2-232.

Listed below are "Cadet" model 111-C entries which constitute the total production of this series:

NC-490K; Model 111-C (# 1) Warner 110
NC-491K;   „   „   (# 2)   „   „

Fig. 95. Nicholas-Beazley NB-3V was "Barling" monoplane with Velie M-5 engine.

The Nicholas-Beazley "Barling NB-3" series were very unusual little monoplanes that incorporated several revolutionary ideas in aircraft construction and aerodynamic arrangement; the one fact that they could carry 3 people with good performance on some 60 h.p., was worthy of some applause in itself. The distinctive design by Walter Barling was indeed a refreshing and somewhat new approach to the never-ending problem of marketing a useful light-plane for personal use, or for fixed-base operations that would show a profit. The NB-3 was certainly economical and efficient in its service, and delivered a performance far beyond the average with this amount of power; its exceptional performance was many times proven by its record. Designed with a simple yet sturdy construction that would render care-free service for many years, with an economy of operation that was hard to beat or even equal per seat-mile, the NB-3 posed as a very attractive buy; because of this, they had sold in good number.

Because of the fact that the Velie M-5 had a good reputation and was such a popular engine, Nicholas-Beazley toyed with the idea of developing this new version of the NB-3 to create a few more sales; the model NB-3V as shown, was the final result. The model NB-3V was also a 3 place open cockpit low-winged monoplane and typical to the earlier NB-3 in most all respects, except for its engine installation; this installation was the 5 cyl. Velie M-5 engine of 55-65 h.p. at 1850-2000 r.p.m. Performance was pretty much comparable with the Velie M-5 and flight characteristics of the new model were also obedient and gentle; a preference for the Velie engine would be the only sensible criteria used in selecting the model NB-3V over the LeBlond-powered NB-3. As an incentive to further sales, Nicholas-Beazley airplanes could now be purchased for a one-third down payment with the balance to be paid up in 12 monthly installments; many of the NB-3 were known to have paid for themselves in this length of time. The type certificate number for the model NB-3V as powered with the Velie engine was issued 9-18-29 and some 21 or more examples of this model were manufactured by the Nicholas-Beazley Airplane Co. at Marshall, Mo.

Listed below are specifications and performance data for the model NB-3V as powered with the Velie M-5 engine; length overall 22'3"; hite overall 7'0"; wing span 32'9"; wing chord 62"; total wing area 160 sq.ft.; airfoil Barling 90-A; wt. empty 772; useful load 629; payload with 18 gal. fuel was 344 lbs.; gross wt. 1401 lbs.; max. speed 100; cruising speed 85; landing speed 38; climb

595 ft. first min. at sea level; climb to 12,000 ft. in 45 min.; ceiling 12,500 ft.; gas cap. 18 gal.; oil cap. 1.5 gal.; cruising range at 4.5 gal. per hour was slightly over 300 miles; price at the factory was $3600.

The fuselage framework was built up of welded chrome-moly steel tubing and streamlined to shape with aluminum alloy fairing structures that faired the fuselage junction into the wing attach point, and also formed the cockpit cowling and the turtle-back section; the completed framework was fabric covered. The 18 gal. fuel tank was mounted high in the fuselage just ahead of the front cockpit, and was actually shaped to form the cockpit cowling as it faired back from the engine compartment. The all-metal wing framework of cantilever design, was built up of box-type spar beams that were formed to shape from stamped-out U-sections of aluminum alloy sheet that were riveted together into spar beams; the wing ribs were riveted together of aluminum alloy sheet stampings. The leading edge of the wing was covered with aluminum alloy sheet to preserve the airfoil form and the completed framework was covered with fabric. This wing structure was quite an innovation at this particular time, and weighed only 1.06 lbs per sq.ft., which was about half the weight for other metal-wing designs of comparable strength. Another useful feature was the ability to procure prefabricated sections of the wing to replace damaged areas; swept-up wing tip sections were also detachable and this prompted at least one clipped-wing version that was used for racing. The fabric covered tail-group was built up of welded chrome-moly steel tubing; the vertical fin and horizontal stabilizer were ground adjustable for trim. The split-axle landing gear used rubber shock-cord (bungee cord) to absorb the bumps; wheels were 24x4 and no brakes were provided. The tail-skid was of the spring-leaf type, with a removable hardened shoe. The next development in the "Barling NB-3" monoplane series was the NB-3G as powered with the 80 h.p. "Genet" engine; see chapter for ATC # 231 in this volume.

Listed below are NB-3V entries that were gleaned from registration records:

NC-884H;	NB-3V	(# 25)	Velie M-5
NC-887H;	,,	(# 28)	,,
NC-508M;	,,	(# 37)	,,
NC-509M;	,,	(# 38)	,,
NC-554M;	,,	(# 47)	,,
NC-555M;	,,	(# 48)	,,
NC-590M;	,,	(# 49)	,,
NC-881M;	,,	(# 53)	,,
NC-884M;	,,	(# 56)	,,
NC-889M;	,,	(# 61)	,,
NC-892M;	,,	(# 64)	,,
NC-894M;	,,	(# 66)	,,
NC-897M;	,,	(# 69)	,,
NC-115N;	,,	(# 73)	,,
NC-116N;	,,	(# 74)	,,
NC-118N;	,,	(# 76)	,,
NC-119N;	,,	(# 77)	,,
NC-120N;	,,	(# 78)	,,
NC-126N;	,,	(# 84)	,,
NC-902N;	,,	(# 88)	,,
NC-132N;	,,	(# 90)	,,
NC-416V;	,,	(# 93)	,,
NC-425V;	,,	(# 100)	,,

Serial # 25 and # 28 first had LeBlond 60; serial # 64 first had LeBlond 60; serial # 76 later had Genet 80 as NB-3G; a few serial numbers have been unaccounted for, these would have been either models NB-3V or NB-3G.

Fig. 96. Unusual lines of NB-3V designed for weight reduction.

Fig. 97. Ability to carry 3 on 65 H.P. very commendable.

Fig. 98. Nicholas-Beazley NB-3G with 80 H.P. British "Genet" engine.

To compliment the NB-3 line of all-purpose monoplanes, Nicholas-Beazley developed the model NB-3G; a worthy addition to the series which offered much in the way of performance and made a name for itself at least on one occasion. Piloted by the veteran D. S. "Barney" Zimmerly, a craft of this type was nursed to over 25,000 feet as a new altitude record for light airplanes. This model was the best performer of the NB-3 series that were in production at this time, and perhaps the best dollar-value of them all, but it did not overtake the other models in popularity. The model NB-3G was the only airplane to be approved on an ATC certificate with the Armstrong-Siddeley "Genet" engine during this period; there were only a few other manufacturers that ventured to use this powerplant. To name some of the others, a Star "Cavalier" model D, a St. Louis "Cardinal", and an improved NB-4G were certificated on Group 2 approvals; one "Monocoupe" was powered with the "Genet" 80 for test. The Fairchild Airplane & Engine Corp. introduced the "Genet" engine in this country, primarily for use in their model "21" which was a low-winged open cockpit monoplane seating two; several of the model "21" were built but a lack of clamorous acceptance caused further development to be discouraged and discontinued. Thus, it remained for Nicholas-Beazley to carry on as the only airplane that had any measure of success with the "Genet" engine at this time in this country.

The Armstrong-Siddely "Genet" Mark 2 of British design and manufacture, was a compact 5 cyl. air-cooled radial engine designed especially for use in light commercial airplanes. Rated at 80 h.p. at 2200 r.p.m., the "Genet" operated at higher shaft speed than the average American engine because of its square (4″ x 4″) bore and stroke. For ease of inspection and maintenance, the two magnetos were mounted on the front crankcase and driven by a cross-shaft, very typical to the arrangement on the Wright "Whirlwind" J4 and J5. The "Genet's" dry weight without propeller hub or starter was 210 lbs. and quoted price at the distributing station was $1500. First imported to this country by the Fairchild Engine Corp. of Farmingdale, L.I., N.Y. for planned use on the "Fairchild 21" series, it was also to be manufactured in quantity by Fairchild in this country; no great demand for this engine developed so the project was discontinued.

The Nicholas-Beazley model NB-3G as shown, was a 3 place open cockpit low-winged monoplane, typical to other models in the NB-3 series except for its powerplant installation; as already mentioned, this was the A-S "Genet" engine of 80 h.p. A performance increase with the added power was quite noticeable, especially on take-off and climb-out, with a few miles added to the overall speed range; otherwise, the NB-3G shared all of the other features which made the NB-3 monoplane a country-wide favorite. The type certificate number for the model NB-3G as powered with the 80 h.p. "Genet" engine was issued 9-18-29 and some 20 or more examples of this model were manufactured by the Nicholas-Beazley Airplane Co. at Marshall, Missouri. The company now had a working force of some 150 craftsmen of various sorts and production had reached one completed airplane per day. Besides operating the nationally-known "airplane parts" supply house, and the airplane manufacturing division, Nicholas-Beazley also had major stock interest in a propeller company and a flying school. Russel Nicholas was the president; Charles M. Buckner was V.P.; Howard Beazley was secretary; Jack Whitaker was sales manager; and T. A. Kirkup became chief engineer after the venerable Walter Barling moved on to other interests and other developments.

Listed below are specifications and performance data for the Nicholas-Beazley model NB-3G as powered with the 80 h.p. A-S "Genet" engine; length overall 22'5"; hite overall 7'0"; wing span 32'9"; wing chord 62"; total wing area 160 sq.ft.; airfoil "Barling 90-A"; wt. empty 735; useful load 629; payload with 18 gal. fuel was 344 lbs.; gross wt. 1364 lbs.; max. speed 105; cruising speed 90; landing speed 38; climb 750 ft. first min. at sea level; ceiling 17,500 ft.; gas cap. 18 gal.; oil cap. 1.5 gal.; cruising range at 4.5 gal. per hour was 340 miles; price at the factory was set for $4050., but later reduced to $3900. Weight figures listed above pertain to model not fitted with oil-cooling radiator; figures for model equipped with oil-cooling radiator are as follows; wt. empty 768; useful load 637; payload with 18 gal. fuel was 352 lbs.; gross wt. 1405 lbs.; oil cap. 2.5 gal.; performance figures were more or less typical.

Construction details of the model NB-3G were typical to all other models in the NB-3 series, therefore we suggest reference to the chapters for ATC # 174 and # 230, for discussion of basic construction and arrangement in detail. The only outstanding difference between the NB-3G and other models of this series, was the provision to mount an oil-cooling radiator to allow higher cruising r.p.m. of the "Genet" engine for longer

Fig. 99. D. S. "Barney" Zimmerley test pilot, flew NB-3G to nearly 25,000 feet.

periods of time. The next developments in the Nicholas-Beazley monoplane were the models NB-4L and NB-4W which will be discussed in the chapters for ATC # 385 and # 386.

Listed below are NB-3G entries that were gleaned from various registration records:

NC-510M;	NB-3G	(# 39)	Genet 80
NC-511M;	„	(# 40)	„
NC-552M;	„	(# 45)	„
NC-590M;	„	(# 49)	„
NC-557M;	„	(# 50)	„
NC-882M;	„	(# 54)	„
NC-896M;	„	(# 68)	„
NC-113N;	„	(# 71)	„
NC-117N;	„	(# 75)	„
NC-118N;	„	(# 76)	„
NC-121N;	„	(# 79)	„
NC-125N;	„	(# 83)	„
NC-127N;	„	(# 85)	„
NC-128N;	„	(# 86)	„
NC-904N;	„	(# 91)	„
NC-556M;	„	(# 92)	„
NC-414V;	„	(# 96)	„
NC-415V;	„	(# 97)	„

Serial # 49 first had Velie as NB-3V; serial # 76 first had Velie as NB-3V; serial # 87 and # 89 later had Lambert 90 as NB-3L; NR-355 (serial # 1) later modified to NB-3W with Warner 90, first had Anzani 60; serial # 3 first had Anzani 60, serial # 4 first had Superior Radial, both modified to NB-3G; NR-546 (serial # 6) first had LeBlond 60, later modified to NB-3G; X-9314 (serial # 16) as NB-3P with Pobjoy engine, tested in service at Marshall Flying School; serial # 80 (-122N) modified to NB-3W with Warner 90, also rumored to have been tested with Warner 110, but this has not been verified; NC-358 (serial # 4) also converted to NB-3G; about 15 of the NB-3 series monoplanes were still in active service some 10 years later.

Fig. 99A. This NB-3G was about the last in the series;
note change-over to stiff legged landing gear and low pressure tires.

A.T.C. #232
(9-18-29)
REARWIN "KEN ROYCE", 2000-C

Fig. 100. Rearwin "Ken-Royce" model 2000-C with Curtiss Challenger engine.

R. A. Rearwin had been well known in the lumber, coal, and petroleum products business in the Salina, Kansas area for a good number of years; although quite interested in aviation for some years, Rearwin didn't become active in the airplane business until June of 1928. Becoming interested and thoroughly convinced in the attractive possibilities in aircraft manufacture, he engaged the services of Fred Landgraf, formerly with the Travel Air Company, and experimental work was soon started on a high-performance biplane of the open cockpit type; a design that was to be a refinement of existing practice, rather than a radical departure from the conventional. The early engineering staff consisted of Landgraf as chief engineer, assisted by J. J. Clark and Wm. Guselman. The first airplane (X-44E) was designed and built in record time and was exhibited proudly at the Chicago Air Show in December of 1928. Somewhat at a loss to come up with a name for their new craft, Rearwin decided upon "Ken-Royce" after his two sons named Kenneth and Royce; thus evolved a catchy name that was

familiar to the industry for many years. The original "Ken-Royce" biplane was built in Salina, Kansas and it was planned to erect a factory for its manufacture in this city, but a 2½ acre site was later constructed on Fairfax Field in Kansas City.

The "Ken-Royce" biplane was a fairly large good-looking airplane somewhat reminiscent of the "Eaglerock" in aerodynamic arrangement, that is to say, it was arranged with long moment-arms, and an abundance of efficient wing area with plenty of interplane gap between panels. Rightfully proud of his new craft, Rearwin entered the "Ken-Royce" in various air-races and air-derbies to show it off. In the Miami to Cleveland Air Derby of 1929, a "Ken-Royce" piloted expertly by George E. Halsey won first in class; in the Woman's Air Derby of 1929 from Santa Monica, Calif. to Cleveland, O., Ruth Nichols was holding a comfortable third in fast company, and then had the misfortune to crack the ship up at Columbus, Ohio. In a closed-course event at the Cleveland Air Races of 1929,

Fig. 101. Aerodynamic arrangement of "Ken-Royce" offered high performance on 175 H.P.

Halsey averaged 124 m.p.h. for 50 miles; in another event of 75 miles, he averaged better than 121 m.p.h. Ruth Nichols, slightly shaken from her previous crack-up, but losing no faith in the ship she had been flying, flew the "Ken-Royce" that Halsey had been flying, in a special event for woman pilots; it is admirable that she finished in 5th spot, considering that all contestants ahead of her had 200 h.p. or better against her 170 h.p. Behavior in competition helped bear out the fact that the "Ken-Royce" biplane was a good performer and quite nimble in relation to its size and the power available.

The Rearwin "Ken-Royce" as shown, was a 3 place open cockpit biplane of rather large proportion but of good aerodynamic arrangement, that was especially planned for the busy flying-salesman or the sportsman-pilot, who would be making frequent short hops in and out of all sorts of landing fields. Powered by the 6 cyl. Curtiss "Challenger" engine of 170 h.p., the "Ken-Royce" offered utility and good performance beyond the average for this amount of power, and consequently was a very good value for the money. Scattered reports conveyed the feeling that flight characteristics were pleasant and control was brisk and sharp; its effervescent personality must have been very enjoyable.

Many have expressed it a down-right shame and pity that this delightful machine was built in such small number. The Rearwin "Ken-Royce" model 2000-C as powered with

the Curtiss "Challenger" engine, received its type certificate number on 9-18-29 and 3 examples of this model were manufactured by the Rearwin Airplanes, Inc. at Fairfax Airport in Kansas City, Kansas. R. A. Rearwin was president; Fred Landgraf was chief engineer; and enthusiastic George E. Halsey was engaged for sales promotion.

Listed below are specifications and performance data for the Rearwin "Ken-Royce" model 2000-C as powered with the 170 h.p. Curtiss "Challenger" engine; length overall 25'0"; hite overall 9'11"; wing span upper 35'0"; wing span lower 31'6"; wing chord upper 66"; wing chord lower 48"; wing area upper 186 sq.ft.; wing area lower 114 sq.ft.; total wing area 300 sq.ft.; airfoil "Rhode-St. Genese"; wt. empty 1495; useful load 885; payload with 55 gal. fuel was 358 lbs.; gross wt. 2380 lbs.; max. speed 130; cruising speed 110; landing speed 45; climb 1000 ft. first min. at sea level; climb to 10,000 ft. in 17 min.; ceiling 24,000 ft.; gas cap. 55 gal.; oil cap. 4 gal.; cruising range at 10.5 gal. per hour was 500 miles; price at the factory was first announced at $8000., later reduced to $6750.

The fuselage framework was built up of welded chrome-moly and low carbon steel tubing, heavily faired to shape with fairing strips and fabric covered; a compartment was under the front seat for 15 lbs. of baggage. This compartment was also used for installation of the storage battery; when battery was installed, no baggage was allowed. The fuel supply was carried in two tanks; main fuel tank of 35 gal. capacity was mounted in the fuselage ahead of the front cockpit, and extra fuel was carried in a 20 gal. tank mounted in the center-section panel of the upper wing. The wing framework was built up of laminated spruce spar beams, with wing ribs of basswood webs and spruce capstrips; the leading edges were covered with birch plywood to preserve the airfoil form and the completed framework was fabric covered. The split-axle landing gear employed Rearwin "rubber and oil" shock absorbing struts; wheels were 28x4 or 30x5 and Bendix brakes were standard equipment. The fabric covered tail-group was built up of welded steel tubing and channel sections; the horizontal stabilizer was adjustable in flight. The elevators were operated by push-pull tubes, the rudder by braided steel cable, and the Friese type ailerons by a combination of torque tubes and flexible steel cable; dual controls were optional. A tail-wheel, navigation lights, and a metal propeller were standard equipment; an inertia-type engine starter was optional. The next development in the "Ken-Royce" biplane was the model 2000-CO as powered with the 165 h.p. Continental A-70 engine; this craft will be discussed in the chapter for ATC # 314.

Listed below are the only examples of the "Ken-Royce" 2000-C that were built:

X-44E; 2000-C (# 101) Challenger 170.
NC-591H;   „   (# 102)   „
NC-592H;   „   (# 103)   „

A group 2 approval numbered 2-106 was issued 8-9-29 for serial # 101 and upwards this approval was superseded by ATC # 232.

Fig. 102. "Ken-Royce" 2000-C used in weather research.

Fig. 103. Command-Aire 5C3-C with 5 cyl. Wright J6 engine, shown at factory.

From all indications that were then present, there is certainly no doubt that the Wright-powered model 5C3-C biplane had a very good chance of becoming the most popular version in the "Command-Aire" 5C3 series. It is a matter of record that Wright engines had an enviable reputation for performance and dependability, and the new J6 series engines were the last word in practical aircraft engine design. The 5 cyl. Wright J6 engine of 165 h.p. was in a power range most suited to all-purpose aircraft of this period, and its performance potential was a boon to aircraft operators the country over. For this reason, a teaming up of the "Whirlwind Five" and the "Command-Aire" biplane produced a combination whose personality was highly rated, and one the company was especially proud of. That the 5C3-C failed to sell in any number was only a quirk of fate and the circumstances that befell all manufacturers across the nation.

The model 5C3-C as shown, was also a 3 place open cockpit biplane of the general-purpose type and was typical to other models in the 5C3 series except for its powerplant installation which in this case was the 5 cyl.

Wright J6 series engine of 165 h.p. Performance of this craft was very good and all of the outstanding attributes of the "Command-Aire" biplane seemed to be especially accentuated in this particular model. Interest in this Wright-powered version surged upward almost immediately but the expected back-log of orders failed to materialize, leaving the company with only the consolation that a future for this craft would have been assured, had the market held up to normal proportions. The model 5C3-C (also as 5C3-5) as powered with the 5 cyl. Wright J6 engine was first certificated on a Group 2 approval numbered 2-117 that was issued 9-6-29 for serial # W-92; an approved type certificate was issued 9-18-29 for serial # W-110 and upwards. Some 5 or more examples of this model were manufactured by Command-Aire, Inc. at Little Rock, Arkansas.

Listed below are specifications and performance data for the "Command-Aire" model 5C3-C as powered with the 165 h.p. Wright J6 engine; length overall 24'6; hite overall 8'6"; wing span upper and lower 31'5"; wing chord both 60"; wing area upper 157 sq.ft.; wing area lower 146 sq.ft.; total wing area 303 sq.ft.; airfoil "Aeromarine" 2A; wt. empty 1559; useful load 931; payload with 55 gal. fuel was 396 lbs.; gross wt. 2490 lbs.; max. speed 120; cruising speed 101; landing speed 42; climb 810 ft. first min. at sea level; ceiling 13,500 ft.; gas cap. 55 gal.; oil cap. 5 gal.; cruising range at 8 gal. per hour was up to 600 miles; price at the factory was $7000., later reduced to $6025.

Construction details and other arrangements typical to most all of the "Command-Aire" biplanes were covered thoroughly in previous chapters (see chapters for ATC # 53 - 118 - 120 - 150 - 151 - 184 - 185 in Vols. 1 and 2 of U.S. CIVIL AIRCRAFT), the following are details pertaining to the model 5C3-C. 30x5 Bendix wheels and brakes were standard equipment and also a ground adjustable metal propeller; a hand-crank inertia-type engine starter and navigation lights were optional equipment. A baggage compartment of 6 cu. ft. capacity was provided for an allowance of 33 lbs.; pilot's seat had deep well for parachute pack with weight allowance of 20 lbs. Development of new models by Command-Aire, Inc. had certainly slowed down by the first part of 1930; the next model brought out was the Warner-powered 2 place trainer (model BS-14) on a Group 2 approval numbered 2-204.

Listed below are model 5C3-C entries that were gleaned from registration records:

NC-932E;	5C3-C	(# W-92)	J6-5-165.
NC-972E;	„	(# W-110)	„
NC-974E;	„	(# W-112)	„
NC-977E;	„	(# W-115)	„
NC-978E;	„	(# W-116)	„

Serial # W-92 was on Group 2 approval numbered 2-117.

Fig 104. 5C3-C was popular for general-purpose work.

Fig. 105. Fokker F-14 transport with 525 H.P. Hornet engine; popular in Canadian service.

The Fokker model F-14 was a large cargo-hauling "parasol" monoplane, the design for which was suggested by veteran Western Air Express pilots; most of the W.A.E. pilots still clung to the belief that the pilot had to be separated from his cargo, and in an open cockpit, to do an efficient job of piloting the airplane. It was also their suggestion earlier that brought about the design of the interesting Lockheed "Air Express", and now fostered the design and the development of the parasol-winged Fokker model F-14. Designed to do double duty, the F-14 was either arranged as a single place cargo-carrier with allowance for nearly 1600 lbs. of payload, as a 7 place transport with seating for 6 passengers and a smaller amount of cargo, which usually consisted of mail, or as a 9 place transport with seating for 8 passengers and the pilot, plus only their baggage. A Wright "Cyclone" powered modification of this design was later used by the Army Air Corps as a cargo-carrier, personnel-transport, or air-borne ambulance, as the Y1C-14 and Y1C-15.

The model F-14 as shown, was a rather large and unusual transport monoplane that closely conformed to unmistakeable "Fokker" construction and configuration, except for the wing's placement above the fuselage and the pilot's open cockpit aft of the cabin section. Basically, the F-14 was offered as a cargo-carrier with bare cabin area of 50x50x114 inch dimensions but provisions were also available for two other interior arrangements. One arrangement carried 6 passengers and some 350 lbs. of cargo, and another arrangement carried 8 passengers and only their baggage in coach-type service. The Fokker F-14 was normally powered with a 9 cyl. Pratt & Whitney "Hornet" A engine of 525 h.p., and performance was rather good for a craft of this bulk and size; it was one of the largest single-engined airplanes in the country at this time. At least six model F-14 were exported to Canada and most of the rest were in use by Western Air Express and later by Trans-continental-Western Air Lines. The type certificate number for the Fokker model F-14 as powered with the 525 h.p. "Hornet" engine was issued 9-21-29 and some 12 or more examples of this model were manufactured by the Fokker Aircraft Corp. of America at Teterboro Airport, Hasbrouck Hts., New Jersey. Harris M. "Pop" Hanshue was president; Wm. T. Whalen was V.P. and general manager; and A. H. G. "Tony" Fokker was chief of engineering. During this particular period, the General Motors Corp. had acquired 40 per cent of Fokker stock; Fred J. Fisher, C. F. Kettering, and C. E. Wilson, all of General Motors, were on the board of directors.

Listed below are specifications and performance data for the Fokker model F-14 as powered with the "Hornet" A of 525 h.p.; length overall 43'4"; hite overall 12'11"; wing span 59'5"; total wing area 550 sq.ft.; airfoil "Fokker"; wt. empty 4346; useful load 2854; payload with 180 gal. fuel was 1460 lbs; gross wt. 7200 lbs.; max. speed 137; cruising speed 116; landing speed 55; climb 810 ft. first min. at sea level; climb in 10 min. was 6000 ft.; ceiling 14,500 ft.; gas cap. 180 gal.; oil cap. 15 gal.; cruising range at 27 gal. per hour was 690 miles; price at the factory was $26,500., later reduced to $22,500; the following figures are for single-place cargo-carrier; wt. empty 4245; useful load 2955; payload with 180 gal. fuel was 1560 lbs.; gross wt. 7200 lbs.; performance figures remained more or less the same; there was a slight gain in available speed with mounting of speed-ring cowling over engine, wheel fairings over main wheels, and a fairing over the tail-wheel.

The fuselage framework was built up of welded chrome-moly steel tubing, faired to shape with formers and fairing strips and fabric covered. The top half of the fuselage, for its entire length, was covered with corrugated duralumin panels; there was an entry door on the left side for entry into the main cabin, with stowage for baggage and a small amount of cargo in a 75 cu. ft. compartment just behind the engine firewall. The one-piece cantilever wing, perched atop 2 heavy N-type struts, was built up of spruce and birch plywood box-type spar beams, with plywood wing ribs and stringers; the completed framework was covered with plywood veneer. The gravity-feed fuel tank was in the center section of the wing. The long-leg landing gear of 162 inch tread used Gruss oil-draulic shock absorbing struts; wheels were 36x8 and Bendix brakes were standard equipment. The fabric covered tail-group was built up of welded chrome-moly steel tubing; the fin was ground adjustable and the horizontal stabilizer was adjustable in flight. A metal propeller, navigation lights, tail-wheel, and hand-crank inertia-type engine starter were standard equipment. The next Fokker development was the 4-engined F-32 transport as discussed in the chapter for ATC # 281 in this volume.

Listed below are Fokker model F-14 entries that were gleaned from registration records:

NC-150H; Model F-14 (# 1400) Hornet.
  CF-AIG;   „   (# 1401)   „

Fig. 106. Prototype F-14 being readied for test.

Fig. 107. Parasol wing and open cockpit aft suggested by air-mail pilots.

CF-AII;	„	(# 1402)	„	NC-328N;	„	(# 1409)	„
CF-AIH;	„	(# 1403)	„	NC-329N;	„	(# 1410)	„
NC-129M;	„	(# 1404)	„	NC-331N;	„	(# 1411)	„
CF-AIJ;	„	(# 1405)	„	NC-332N;	„	(# 1412)	„
CF-AIK;	„	(# 1406)	„				
CF-AIL;	„	(# 1407)	„				
NC-327N;	„	(# 1408)	„				

Registration number unverified for serial # 1410.

Fig. 108. Army F-14 was YIC-14; another version served as ambulance.

A.T.C. #235
(9-21-29)
MERCURY "CHIC", T-2

Fig. 109. Mercury "Chic" T-2 with 90 H.P. Le Blond engine.

The Mercury "Chic" was a rather homely and somewhat unusual configuration that in appearance belied its actual merit; a sound design incorporating many functional and some out-of-ordinary features. This open cockpit parasol monoplane had seating for two in tandem and its performance on 90 h.p. was quite exceptional; visibility was excellent and its rugged construction afforded reasonable trouble-free operation. Altogether the piquant "Chic" can be summed up as a craft with many advanced and sensible features for the average private owner engaged in sport flying for the fun of it. The Mercury "Chic" as powered with the 7 cyl. LeBlond (7D) engine of 90 h.p. was awarded a Group 2 certificate numbered 2-126 which was superseded by ATC # 235 issued 9-21-29. Coming out in prototype in the latter part of 1928, some 13 or more examples of this model were manufactured and sold by Mercury Aircraft, Inc. at Hammondsport, New York.

In view of the fact that the beginning and end of aircraft manufacture by Mercury Aircraft, Inc. lies in this one chapter, it would be fitting to elaborate somewhat on the company's history. First organized as the Aerial Service Corp. in 1920, its president was Henry Kleckler, a mechanic and engineer for Glenn Curtiss the 13 years previous; with Wm. C. Chadeayne as vice-president. Operation of the new firm was established in a former barrel factory in Hammondsport, N.Y., not too far from the scene of early Curtiss activity; production first consisted mostly of wood and metal airplane and air-ship parts, largely for the government. In 1924 the Aerial Service Corp. was fortunate to acquire the services of two very capable men; Joseph F. Meade and Harvey C. Mummert. These two former Curtiss employees sparked a new life and had a great influence on the future of the company; Joseph Meade became general manager and manager of sales, while Harvey Mummert became the chief engineer to carry on with new aircraft developments. Mummert's design and engineering talents can also be reflected in several early Curtiss aircraft of merit. The first new airplane design undertaken

Fig. 110. Prototype "Chic" with Velie engine; arrangement was unusual but efficient.

at Aerial Service was the "Mercury Sr." biplane, a large "Liberty" powered mail-plane that was purchased in 1925 by the Post Office Dept. for use on eastern routes; the mail-carrying "Mercury" was one of several other designs submitted by various firms for service tests. Another type produced in 1925 was a highly modified Standard J-1 that carried 5 people in open cockpits with the power of a Curtiss C-6A engine; another modified Standard version was a 2-3 place sport-trainer model. The second all-new design was a smaller all-purpose airplane of 3 place seating that was called the "Mercury Jr."; this versatile plane, flown by Harvey Mummert in the 1925 Ford Air Tour with Wm. Chadeayne as passenger, had made a commendable showing in performance, in spite of the troubles encountered enroute. Entered in both the 1926 and 1927 Air Tour, Mummert made a commendable showing again and garnered a third spot in the 1927 event, in spite of the fierce competition. To bolster income, the company also engaged in

development of lighter-than-air vehicles; the TC-11 designed by Mummert was accepted and delivered to the Army in 1928. Late in 1927, a new design for a commercial aircraft was begun; it was to be a 3 place enclosed monoplane called the "Kitten". Of interesting design, the "Kitten" was exhibited hopefully in 1928, was in active service for many years, and was finally scrapped in 1946. Re-organized again with new capital in 1929, the company name was changed to Mercury Aircraft, Inc.; John F. Wentworth became the new president and Joseph Meade and Harvey Mummert became vice-presidents, retaining their position of general manager and chief engineer respectively. Of particular interest was the Schroeder-Wentworth "safety-plane" that was an entrant in the Guggenheim Safe Airplane Contest of 1929; built at Mercury Aircraft, the basic specifications for this airplane were laid-out by Maj. R. W. "Shorty" Schroeder with design details and engineering by Harvey Mummert. Typical of all aircraft designers, Mummert always had new designs in the back

Fig. 111. Parasol wing on "Chic" offered good visibility and pendulum stability.

Fig. 112. 1926 Mercury "Arrow: boasted high performance.

of his head or on the drawing board; in 1928 he had designed the very distinctive "Chic".

The "Chic" was planned as a primary-trainer for operation off small air-fields with good performance and amiable flight characteristics. The prototype "Chic" was powered with a 65 h.p. Velie engine and was fitted with a N.A.C.A. type low-drag engine cowling, which was still unusual on a craft of this type at this time. Performance with the Velie engine was not quite satisfactory so the 7 cyl. LeBlond 7D of 90 h.p. was used in subsequent airplanes. The first few craft of this model had no fixed tail surfaces, both rudder and elevators were of the balanced all-flying type; this promoted somewhat sensitive control and was somewhat hard to get accustomed to. To alleviate this, the tail surfaces were later redesigned to a more conventional type. Visibility and manuverability of the "Chic" were exceptional but the extremely high wing location, coupled with 26x4 tires and no brakes, did give many pilots a few anxious moments in ground maneuvering, or on take-offs and landings in a stiff off-side wind. "Shorty" Schroeder's affiliation with the far-flung Curtiss Flying Service at this time provided a ready-made market for the "Chic" trainer; quite a few were in use for pilot training at Curtiss schools. To help promote sales, Harvey Mummert flew a "Chic" in the National Air Tour for 1930; showing off the plane in 27 cities and finally placing 17th in the point standings. In the 1931 National Air Tour, Mummert flew a "Chic" again but due to misfortune was forced to withdraw.

Production of the amiable "Chic" continued on but the market for airplanes of this type was greatly affected by the financial depression of the early thirties, and consequently quite a few "Chics" remained unsold. In 1931, the Schroeder and Wentworth interests were withdrawn from the company and Joseph Meade with Harvey Mummert took over complete control; Meade was now the president, general manager, and sales manager while Mummert was V.P., secretary-treasurer, chief engineer, and chief pilot. Business had been very slow and the factory was down to four employees; to keep the plant and the employees going, Mummert designed another racing craft similar to the one he had flown in the celebrated "Cirrus Derby". It was hoped that a successful racing airplane might spark

Fig. 113. Mercury "Kitten" was cabin monoplane with Warner engine.

a little interest in the company's products. In 1938 a "Chic" was modified somewhat to meet the requirements of an Air Corps competition for a primary trainer; the "Chic" was termed as unsatisfactory despite its merits, so this venture was the last of "Chic" development. At the end of World War 2, there were still over a dozen of "Chic" T-2 airframes and other parts left over; with storage space at a premium it was decided to scrap all but one, which was ear-marked for exhibition in a proposed air-museum in Hammondsport. Mercury Aircraft, unlike many others, continued a struggle throughout the shaky thirties with spare parts and sub-assemblies on government contract, and thrived on this business through W.W.2 and for the next 20 years. Harvey C. Mummert died after illness in 1939, and Joseph Meade died in 1950; it was left for young Joseph Meade, Jr. to carry on the more than 30 years of determined tradition.

Listed below are specifications and performance data for the "Mercury Chic" model T-2 as powered with the 90 h.p. LeBlond 7D engine; length overall 23'0"; hite overall 8'7"; wing span 35'8"; wing chord 66"; total wing area 192 sq.ft.; airfoil Clark Y; wt. empty 935; useful load 578; payload with 28 gal. fuel was 208 lbs.; gross wt. 1513 lbs.; max. speed 115; cruising speed 95; landing speed 40; climb 900 ft. first min. at sea level; ceiling 18,000 ft.; gas cap. 28 gal.; oil cap. 4 gal.; cruising range at 6.5 gal. per hour was 380 miles; price at the factory was $4250., lowered to $3980.

The fuselage framework was built up of welded chrome-moly and 1025 steel tubing, faired to shape with steel tube fairing strips and the entire fuselage, including cockpit coaming, was fabric covered from the firewall back. The fuselage shape terminated to a knife-edge on the horizontal, to which the balanced elevator was fastened. The wing framework was built up of truss-type spar beams of welded chrome-moly steel tubing, with wing ribs fashioned of welded 1025 steel tubing; full-span ailerons were of the same construction and the complete framework was fabric covered. Wing brace and wing attach struts were of chrome-moly steel tubing in a streamlined section. The gravity-feed fuel tank was mounted high in the fuselage, just ahead of the front cockpit. The fabric covered tail-group was built up of welded chrome-moly steel tubing; the fin was ground adjustable and the all-flying elevators were adjustable in flight. Trim was affected in regulating deflection of the elevators by winding up tension on a bungee-cord servo mechanism. The split-axle landing gear was of chrome-moly steel tubing in streamlined section, with one spool of rubber shock-cord to absorb the bumps; wheels were 26x4 with no brakes, although brakes were available as optional equipment.

Listed below are Mercury "Chic" T-2 entries that were gleaned from various records:

NC-485E;	Chic T-2	(# 20)	Velie M-5.
NC-883K;	„ „	(# 21)	LeBlond 90
NC-814M;	„ „	(# 22)	„
NC-815M;	„ „	(# 23)	„
NC-15N;	„ „	(# 24)	„
NC-16N;	„ „	(# 25)	„
NC-17N;	„ „	(# 26)	„
NC-48N;	„ „	(# 27)	„
NC-49N;	„ „	(# 28)	„
NC-50N;	„ „	(# 29)	„
NC-51N;	„ „	(# 30)	„
NC-52N;	„ „	(# 31)	„
NC-53N:	„ „	(# 32)	„

Serial # 20 prototype, later mounted LeBlond 90; serial # 31 rebuilt for Air Corps trainer competition; serial # 33 through # 39 and # 41 through # 48 were balance of 1929 production run which remained unsold, all were scrapped in 1946; serial # 40 was saved for Hammondsport museum; serial # 49 was "Mercury S" racer; serial # 52 was "Mercury S-1" racer; no record for serial # 50 and # 51.

Fig. 114. Harvey Mummert, designer, flew "Chic" in 1931 National Air Tour.

A.T.C. #236
(9-25-29)
CURTISS "THRUSH", MODEL J

Fig. 115. Curtiss "Thrush" model J with 225 H.P. Wright J6 engine.

Announced as a big-sister version of the popular "Robin", the Curtiss "Thrush" was first introduced late in 1928 in a Curtiss "Challenger" powered version, but exhaustive tests in several configurations had finally proven that this big craft with 170-185 h.p., would certainly be no match for other aircraft of this particular type. The flying and buying public had come to expect certain capabilities by now, and the trend was to higher speeds and more horsepower. Satisfied that the "Challenger" engine would not do the job, Curtiss engineers modified the 3 examples of the "Thrush" that were previously built, with the installation of the 7 cyl. Wright J6 engine of 225 h.p. The installation of the more powerful engine required only little modification to the craft in general, and a test program soon pronounced the 6 place "Thrush" monoplane satisfactory and quite ready for quantity production; consequently, the "Thrush" program was turned over to the Curtiss-Robertson division at St. Louis, which had the proper facilities for its manufacture and tooling-up for the job was accomplished in short order.

The Curtiss "Thrush" as powered with the Wright J6 engine of 225 h.p., was a 6 place high wing cabin monoplane of the boxy lines quite similar to the "Robin", and its performance was now nearly comparable to other airplanes of this size and capacity. Sensibly appointed and roomy, the "Thrush" had capacity for a sizeable payload and short-field performance was one of its better features; this points to its frequent use as an air-taxi that operated into and out of the pasture-airports that were more the rule than the exception, during this period of time. Serving as a shake-down for the new design, a Curtiss "Thrush" model J was flown by Dale "Red" Jackson in the National Air Tour for 1929; another was flown by J. L. McGrady and they garnered a 12th and 18th place in the final standings. The type certificate number for the Curtiss "Thrush" model J, as powered with the Wright J6 (R-760) engine, was issued on 9-25-29 and this specification only applied to two of the 3 examples that were built previously under ATC # 159. (Refer to U.S. CIVIL AIRCRAFT, Vol. 2). The Challenger-powered "Thrush" was also approved earlier to the Curtiss-Robertson division on ATC # 160, but no production was scheduled be-

Fig. 116. Likened to "Big Robin" the "Thrush" was light transport for six.

cause of the early decision to modify the design with more horsepower.

Listed below are specifications and performance data for the Curtiss "Thrush" model J as powered with the 225 h.p. Wright J6 engine; length overall 32'7"; hite overall 9'3"; wing span 48'0"; wing chord 84"; total wing area 305 sq.ft.; airfoil Curtiss C-72; wt. empty 2260; useful load 1540; payload with 60 gal. fuel was 970 lbs.; payload with 110 gal. fuel was 670 lbs.; gross wt. 3800 lbs.; max. speed 122; cruising speed 104; landing speed 52; climb 650 ft. first min. at sea level; climb to 5000 ft. in 9.4 min.; climb in 10 min. was

5300 ft.; ceiling 13,200 ft.; gas cap. normal 60 gal.; gas cap. max. 110 gal.; oil cap. 5-9 gal.; cruising range from 490-900 miles; tentative price at the factory was $12,000.

Following traditional Curtiss practice, the fuselage framework was built up of seamless dural tubing and fastened together with wrap-around fittings and hollow dural rivets; some of the more highly stressed portions of the framework were built up of welded chrome-moly steel tubing. The completed framework was lightly faired to shape and covered with fabric. Cabin appointments were plain and serviceable, with entry gained through two large doors on the right side. The semi-canti-

Fig. 117. "Thrush" refueled in mid-air by "Robin" tanker.

lever wing framework was built up of solid spruce spar beams and stamped-out aluminum alloy wing ribs; fittings were installed to allow folding of the wings to a width of 12.5 feet, and the completed framework was covered with fabric. Fuel tanks were mounted in the wing, with one placed on either side of the fuselage; wing braces were parallel struts with a strut junction cabane similiar to the arrangement used on the "Robin". With seats removed, useable cabin area was 38x41x81 inches. The split-axle landing gear of 116 inch tread employed oleo-spring shock absorbers; wheels were 30x5, Bendix brakes and a full-swivel tail-wheel were provided. The fabric covered tail-group was built up of welded chrome-moly steel tubing and sheet steel formers; the fin was ground adjustable

and the horizontal stabilizer was adjustable in flight. Dual controls, a Curtiss-Reed metal propeller, inertia-type engine starter, and wiring for navigation lights, were standard equipment. The Curtiss-Robertson production version of the "Thrush" model J will be discussed in the chapter for ATC # 261 in this volume.

Listed below are Curtiss "Thrush" model J entries that were converted from two of the three aircraft built previously:
C-7568; Thrush J (# G-1) J6-7-225.
C-9787;   „   (# G-2)   „

Serial # G-3 (NR-9142) was modified into a J Special with installation of the Wright J6-9-300 engine.

Fig. 118. Boxy lines of "Thrush" surprisingly efficient; entry afforded by convenient steps and two large doors.

Fig. 119. Curtiss "Carrier Pigeon" model 2 with 12 cyl. geared "Conqueror" engine of 600 H.P.

In answer to queries for a huge single-engined airplane that could carry a ton or so of payload, Curtiss revived the development of the old "Carrier Pigeon" design. This new development was the model 2 which carried a much larger payload than the old design, at higher cruising speeds, and was by far the biggest and the heaviest single-engined mail-plane of this time, and for at least a year or so to come. The "Carrier Pigeon" model 2 was very much unlike the spirited "Falcon" mail-planes that carried smaller payloads at much higher speeds; the "Carrier Pigeon", outweighing the "Falcons" by more than a ton, was more like an air-borne truck or freighter that was designed for more than twice the payload. Though there was some sacrifice in speed and maneuverability due to its enormous bulk, the "Carrier Pigeon" 2 was a much cleaner design aerodynamically than the "Falcons", which were quite dirty by comparison. Extra fuel load extended the cruising range considerably and this allowed greater intervals between fueling stops. National Air Transport (N.A.T.) had 3 of the "Carrier Pigeon" 2 and they were operated with satisfaction and profit.

The first "Carrier Pigeon" (Model 1), shown here for comparison, was powered with the 12 cyl. war-born "Liberty" engine and was introduced early in 1925; this prototype was flown in the Ford Air Tour for 1925 by the inimitable "Casey" Jones as a shake-down cruise, and the performance shown despite minor mishap was commendable for a craft of this size. This version was the Model 1 and several were operated by N.A.T.; at least 11 were on the flight roster during one particular time. Except for name, there is very little resemblance between the "Carrier Pigeon" models 1 and 2; the new model 2 was powered with a geared-down (2 to 1) 12 cyl. vee-type Curtiss "Conqueror" engine of 600 h.p., and its carrying capacity and operating speeds greatly out-stripped the older design. Designed for durability and efficiency with most of the latest aeronautical developments, the Model 2 provided comforts and conveniences for the pilot, was quite easy to fly and maneuver on the ground, with special attention paid to efficient loading and servicing.

The Curtiss Carrier Pigeon model 2, as powered with the 600 h.p. Curtiss "Conqueror" engine, was awarded its type certificate number on 9-25-29 and 3 examples of this model were manufactured by the Curtiss Aeroplane & Motor Co., Inc. at their plant in Buffalo, New York.

Fig. 120. Unusual view of "Carrier Pigeon" 2.

Listed below are specifications and performance data for the Curtiss "Carrier Pigeon" model 2 as powered with the 600 h.p. "Conqueror" engine; length overall 34'6"; hite overall 13'4"; wing span upper 47'6"; wing span lower 43'6"; wing chord upper 88"; wing chord lower 68"; total wing area 553 sq.ft.; airfoil Curtiss C-72; wt. empty 4210; useful load 3390; payload with 175 gal. fuel was 2022 lbs.; gross wt. 7600 lbs.; max. speed 150; cruising speed 123; landing speed 65; climb 850 ft. first min. at sea level; climb in 10 min. was 6400 ft.; ceiling 12,200 ft.; gas cap. 175 gal.; oil cap. 16 gal.; cruising range at 35 gal. per hour was 580 miles.

The huge fuselage was a composite structure fabricated in 3 separate units then bolted together; the forward unit mounting radiator and engine was a steel tube structure covered with aluminum cowling panels. The center portion up to the pilot's cockpit was a semi-monococque structure of dural formers and stringers, with a riveted dural sheet outer covering that carried a percentage of the stresses; two large cargo holds were in this

portion. The aft section including the pilot's cockpit, was a structure of dural tubing that was riveted together in truss form, faired to shape and fabric covered. The pilot's cockpit was roomy and comfortable, there were several foot-holds and hand-holds for ease of entry, and a heater-muff off the engine's exhaust pipe provided cockpit heat; a baggage compartment for the pilot's personal effects was in the turtle-back section of the fuselage behind the seat. One descriptive account stated that this compartment was for the stowage of the pilot's parachute, but one would think that he'd rather wear it than have it back there in the compartment. The four wing panels were built up of spruce box-type spar beams with spruce and plywood girder-type wing ribs; the ailerons were built up of dural channel-section spars and ribs, the leading edges were covered with dural sheet and the completed framework was fabric covered. A good portion of the fuel was

Fig. 121. 1925 "Carrier Pigeon" model 1 had famous "Liberty" engine.

carried in a belly-tank that was faired into the under-side contours of the fuselage; fuel was pumped into a gravity-feed tank that was mounted in the root end of the left upper wing. Fairing fillets were used at the junction of the fuselage and the lower wing to improve the airflow in this area. The fabric covered tail-group was built up of dural channel-section spars and former ribs; the fin was ground adjustable and the horizontal stabilizer was adjustable in flight. The split-axle landing gear employed "oleo & rubber" shock absorbing struts made by Curtiss; wheels were 36x8 and Bendix brakes were standard equipment. The tunnel-type engine radiator was faired neatly into the forward fuselage contours and had adjustable shutters to control the engine temperature; an Eclipse hand-crank or electric inertia-type engine starter was provided. A large 3-bladed Curtiss-Reed metal propeller, wiring for night flying equipment, including landing lights and parachute flares was standard equipment. The next Curtiss development was the Fledgling J-1 as discussed in the chapter for ATC # 266 in this volume.

Listed below are Curtiss "Carrier Pigeon" Model 2 entries that were gleaned from various records:

NC-985H; Model 2 (# G-1) Conqueror 600.
NC-311N;   „   (# G-2)   „
NC-369N;   „   (# G-3)   „

Fig. 122. Handsome "Pigeon Two" carried over one ton of mail and cargo.

A.T.C. #238
(9-26-29)
SIMPLEX "RED ARROW", W2S

Fig. 123. Simplex "Red Arrow" W2S with 110 H.P. Warner engine.

Though somewhat rare and scarce in numbers built, the Simplex "Red Arrow" mid-wing monoplane was quite well known in flying circles the country over. It cannot be truthfully said this saucy little craft was extremely popular, but it was a welcome visitor to any airport, and always the subject of interesting discussion. Being of mid-wing configuration, the "Red Arrow" monoplane was shied away from by a good many, because as yet, the majority seemed to believe that the monoplane wing belonged on top; as a result, many wild and untrue stories were circulated about "the tricky mid-wing". The Simplex "Red Arrow" was still the only certificated mid-wing monoplane during this period and remained the only one for a year or so more. In 1930-1931, the "Emsco" mid-wing sport-trainer monoplane was finally approved, but it did not enjoy much success and we might say that the Buhl "Bull Pup" was perhaps the only mid-wing monoplane that enjoyed a good measure of popularity for any length of time.

The Simplex "Red Arrow" model W2S as shown and discussed here, was a progressive development of this design and though quite typical, the W2S had several airframe improvements to offer. Powered with the 7 cyl. Warner "Scarab" engine of 110 h.p., performance of this craft was quite snappy and its eager nature was well suited to the requirements of the low-budget sportsman pilot.

Seating two side by side in a fairly roomy open cockpit, the "Red Arrow" W2S could be quite docile and well-mannered, and was often used for pilot training in primary or secondary phases of instruction. The mid-wing placement had a natural tendency to blank off a direct view downward, but visibility above, to the rear, and in turns, was excellent. Maneuverability of the frisky W2S was sharp and precise as demonstrated time and again by the inimitable H. S. "Dick" Myrhes, the pilot who used to "scare the pants off people" with his near-reckless flying and devil-may-care abandon. The Simplex "Red Arrow" model W2S as powered with the 110 h.p. Warner engine, received its type certificate number on 9-26-29 and some 6 or more examples of this model were manufactured by the Simplex Aircraft Corp. at Defiance, Ohio. E. J. Allen was president; F. W. Allen was V.P.; G. H. Roberts was secretary-treasurer and sales manager; O. L. Woodson was chief of engineering; and F. H. Greime, former test-pilot at Simplex, was replaced by H. S. Myrhes who became chief pilot in charge of test and promotion.

The Simplex Aircraft Corp. was always an enterprise of quite modest proportion, with a working force at most of some 25 or 30 people who were craftsmen of all sorts; they also had ample facilities for their modest production in a plant well situated next to a rail-siding. Later on, they had promoted an ad-

*Fig. 124. Simplex convertible mono-biplane;
shown here with bottom wing removed.*

joining air-field. In 1929, Simplex developed
an unusual convertible airplane that could be
flown either as a biplane or a monoplane.
The extra lifting area of the biplane was to
provide the ability to carry heavier pay-loads,
and the conversion of the biplane into a
monoplane (simply by removing the lower
wings), was to provide greater cruising speeds
and range for cross-country travel. This was
truly a sportsman-pilot airplane and only one
example was built. Flown by "Dick" Myrhes
in several events of the 1929 National Air
Races, this craft as a monoplane averaged up
to 152 m.p.h. around the pylons on 225 h.p.
In late 1930 the sales for airplanes was falling
off to nothing and Simplex Aircraft decided
to enter the fracas with an ultra-light airplane; most aircraft manufacturers now felt
that the cheap-to-operate ultra-lite airplane
was destined to revive aircraft buying back
to comfortable levels. The ultra-light offering
by Simplex was the peculiar "Kite", as shown,
but it must have been built only in a prototype. Shortly thereafter, O. L. Woodson and

H. S. Myrhes left for the west coast where they
developed the "Cycloplane". The "Cycloplane" was offered as a penguin-type trainer
for ground maneuvering, and also as an airplane; both models were powered with the
"Cyclomotor". For reference and comparison
of the earlier Simplex K2S and K2C, against
the W2S, refer to U.S. CIVIL AIRCRAFT,
Vol. 1 in the chapters for ATC # 43 and
44.

Listed below are specifications and performance data for the Simplex "Red Arrow"
model W2S as powered with the 110 h.p.
Warner engine; length overall 22'2"; hite
overall 6'11"; wing span 34'4"; wing chord
60"; total wing area 158 sq.ft.; airfoil Clark
Y; wt. empty 1152; useful load 627; payload
with 44 gal. fuel was 174 lbs.; gross wt. 1779
lbs.; max. speed 125; cruising speed 107;
landing speed 45; climb 900 ft. first min. at
sea level; ceiling 12,000 ft.; gas cap. 44 gal.;
oil cap. 3 gal.; cruising range at 6.5 gal. per
hour was 650 miles; price at the factory was
$5500., later reduced to $4495.

The fuselage framework was built up of
welded chrome-moly steel tubing, faired to
shape with wooden fairing strips and fabric
covered. The side-by-side seats were of wicker
and upholstered in leather; there was a
baggage compartment to the rear of the seats
but no baggage was allowed with a full gross
load. The wing framework was built up of
spruce and plywood box-type spar beams,
with spruce and basswood truss-type wing

Fig. 125. "Red Arrow" in package express service at Ford airport.

Fig. 126. Revealing view showing bracing, landing gear arrangement and visibility of "Red Arrow."

ribs; the ailerons were a framework of welded steel tubing, the wing leading edges were covered with dural sheet, and the completed framework was covered in fabric. Two wing stubs projected downward from the lower fuselage to provide attach points for the wing bracing and also for the landing gear; this was a somewhat simpler system than used on the earlier K2S and K2C. The wing stubs, landing gear, and wing bracing struts, were fabricated from chrome-moly steel tubing. Hand-holds were provided in each wing tip and steps were provided in the fuselage for entry to the cockpit. The wide-tread landing gear used oil-draulic shock absorbing struts; wheels were 26x4 and brakes were optional. The fabric covered tail-group was built up of welded chrome-moly steel tubing; the fin was ground adjustable and the horizontal stabilizer was adjustable in flight. A wooden propeller, navigation lights, log books, and fire extinguisher were standard equipment. A metal propeller, dual controls, inertia-type engine starter, wheel brakes, and low pressure "airwheels" were offered as optional equipment. Some of the later examples of this model were eligible with the 125 h.p. Warner engine. After floundering for a while, Simplex Aircraft finally went into receiver-ship in

July of 1932; the untimely end of a promising enterprise.

Listed below are Simplex model W2S entries that were gleaned from various registration records:

X-9412;	Model W2S	(# 28)	Warner-110
NC-371V;	,,	(# 30)	,,
NC-228K;	,,	(# 35)	,,
NC-198N;	,,	(# 36)	,,
NC-957W;	,,	(# 39)	,,
-153Y;	,,	(#)	,,

Serial # 28 modified from K2S; registration numbers for serial numbers 38-40-41-42 unknown; model R2D convertible monobiplane was NR-43M (serial # 37-R); model S-2 "Kite" was X-489M (serial # 1002); there may have been another "Kite" built, but available records do not show it.

Fig. 127. Simplex "Kite" was ultra-light experiment.

A.T.C. #239
(9-26-29)
BRUNNER-WINKLE "BIRD", BK

Fig. 128. Bird BK with 100 H.P. Kinner engine; flown by Lee Gelbach in 1931 National Air Tour.

The trim and graceful "Bird" biplane by Brunner-Winkle had made a big hit with the flying populace and was selling a fairly good number; powered with the 8 cyl. Curtiss OX-5 engine of 90 h.p., the Model A version was actually curtailed in potential number because of the short supply of OX-5 engines. Stock-piles of this war-surplus engine were finally being depleted and new engines were still hard to find; rebuilt engines were used in some cases as a last resort and very often, the customer was obliged to furnish his own engine for installation at the factory. Sensing the inevitable, Brunner-Winkle made plans to standardize production of the "Bird" biplane with a small radial air-cooled engine which would offer comparable performance. In October of 1929 it was formally announced that the 5 cyl. Kinner K5 engine of 100 h.p. would be the standard powerplant installation for the new "Bird" model B. The new model B prototype, as shown here, was introduced for test about the first of the year and it proved to be quite a sensation. Performance was increased to some extent and all the other amiable characteristics of the "Bird" design were retained and in many instances im-

proved. Production of the Model B got off to a slow start because of the business slump in the latter part of 1929 and early 1930, but sales picked up gradually and soon the new "B" was gracing airports all over the country. The rugged character and the inherent safety of this new version naturally promoted longevity, so it is not surprising that at least 60 were still in active operation some 10 years later. The model BK has been a much-sought airplane for restoration, even in the 1960 period.

The new Brunner-Winkle "Bird" model B, later redesignated to BK to identify installation of the Kinner K5 engine, was also a 3 place open cockpit biplane of the sesqui-wing arrangement that offered advantages in aerodynamic efficiency and structural rigidity. Well planned for the sort of utility work that was expected from a general-purpose airplane, the "Bird" was convenient, gentle, honest, and completely trustworthy; its nature was inclined to give the pilot ample time for decision in case of emergency. The BK was especially nice to the lady-pilots and they loved it. In shopping around for a suitable airplane, Chas. A. Lindbergh, famous New

York to Paris flyer, put the "Bird" BK through the paces, as he well could do, and thought so well of its personality and ability that he bought one for his beloved wife Anne to fly around.

The "Bird" biplane was capable of outstanding slow-speed performance with quick and short take-offs, good rate of climb-out, and unbelieveably short landings; a performance that approached that of an autogiro. An unconfirmed story relates that a "Bird" biplane actually bested an autogiro during some unofficial tests; true or not, with a good pilot aboard it was entirely possible. The "Bird" BK was certainly capable of a fair top speed but it always looked like it was going slow, perhaps because of its sesquiplane arrangement that created an illusion; to emphasize this illusion is a story. This story relates of two airmen talking and one says, "There's a Bird coming in". "Coming in?" — "Man, he's buzzing the field!" — flips the other.

The Brunner-Winkle "Bird" model BK, as powered with the 100 h.p. Kinner K5 engine, received its type certificate number on 9-26-29 and some 75 or more examples of this model were manufactured by the Brunner-Winkle Aircraft Corp. at Glendale, Brooklyn, New York. William E. Winkle was president; A. Brunner was secretary; and Michael Gregor was chief of engineering. In 1930 the firm was feeling a dire financial pinch and was reorganized with new capital from Detroit as the Bird Aircraft Corp., with Maj. Thomas Lanphier, former commandant of the First Pursuit Group, as the president; Wm. E. Winkle was V.P.; James Phelan was secretary

and Michael Gregor stayed on for a time as chief engineer. Reference is made here to U.S. CIVIL AIRCRAFT, Vol. 2 in the chapter for ATC # 101, where it was stated that we were not sure of Mr. Brunner's official function in the Brunner-Winkle firm; it has since been established that A. Brunner was the secretary and also provided a share of the financial backing. It was also stated in this chapter that we believed Michael Gregor was the designer of the "Bird" biplane series, but had no substantiating proof; it has since been established that Michael Gregor definitely was the designer of the "Bird" biplane series and also the chief engineer for the Brunner-Winkle firm.

Listed below are specifications and performance data for the "Bird" model BK as powered with the 100 h.p. Kinner K5 engine; length overall 23'0"; hite overall 8'8"; wing span upper 34'0"; wing span lower 25'0"; wing chord upper 69;" wing chord lower 48"; wing area upper 184 sq.ft.; wing area lower 82 sq.ft.; total wing area 266 sq.ft.; airfoil USA-40B modified; wt. empty 1199; useful load 781; payload with 37 gal. fuel was 374 lbs.; gross wt. 1980 lbs.; max. speed 110; cruising speed 92; landing speed 35; climb 650 ft. first min. at sea level; ceiling 14,000 ft.; gas cap. 37 gal.; oil cap. 2.5 gal.; cruising range at 6 gal. per hour was 500 miles; price at the factory was $4995., lowered to $4095., and cut to $3895. in late 1930.

The fuselage framework was built up of welded chrome-moly steel tubing in Warren truss form, with all fittings slotted and welded into the fuselage tubes; the completed framework was faired to shape with wooden

Fig. 129. Prototype Bird BK shows Sesqui-Wing arrangement.

fairing strips and fabric covered. The entire top section of the fuselage was covered with removable aluminum panels; the metal turtle-back section contained a baggage compartment with allowance for 30 lbs., and was quickly removable for inspection or maintenance of the control operating mechanisms in the aft fuselage. The wing panels were built up of heavy spruce spar beams of solid section, with spruce and plywood truss-type wing ribs; the leading edges were covered with duralumin sheet and the completed framework was covered with fabric. The small lower wings and the large amount of stagger between panels enabled the pilot to see his wheels at all times; this a feature quite helpful during take-off and landing, and very helpful in ground maneuvering. The split-axle landing gear was of the cross-axle type using oil and rubber shock absorbing struts; the tail-skid was of the simple spring-leaf type. The model BK was also eligible to operate on skis. The main fuel tank was in the forward section of the fuselage ahead of the front seat; the front joy-stick was removable when carrying passengers. The fabric covered tail-group was built up of welded chrome-moly steel tubing; the fin was ground adjustable and the horizontal stabilizer was adjustable in flight. A metal propeller, Heywood air-operated engine starter, navigation lights, and wheel brakes, were

Fig. 130. "Kinner-Bird" was very popular for general-purpose work; performance was tops.

offered as optional equipment. The next development in the "Bird" biplane was the Warner-powered model BW as discussed in the chapter for ATC # 382.

Listed below are "Bird" BK entries that were gleaned from registration records; a complete listing would be too extensive, so we will list the first 20 or so:

X-221E;	Bird BK	(# 2000)	Kinner K5
NC-77K;	,,	(# 2002-53)	,,
NC-44K;	,,	(# 2003-54)	,,
NC-47K;	,,	(# 2004-55)	,,
NC-81K;	,,	(# 2005-56)	,,
NC-48K;	,,	(# 2006-57)	,,
NC-45K;	,,	(# 2007-58)	,,
NC-18K;	,,	(# 2008-)	,,
NC-79K;	,,	(# 2010-69)	,,
NC-82K;	,,	(# 2011-70)	,,
NC-80K;	,,	(# 2012-73)	,,
NC-211V;	,,	(# 2013-97)	,,
NC-213V;	,,	(# 2014-02)	,,
NC-212V;	,,	(# 2015-01)	,,
NC-946V;	,,	(# 2016-68)	,,
NC-944V;	,,	(# 2017-03)	,,
NC-972V;	,,	(# 2018-99)	,,
NC-974V;	,,	(# 2019-98)	,,
NC-977V;	,,	(# 2020-05)	,,
NC-980V;	,,	(# 2021-95)	,,
NC-812W;	,,	(# 2022-00)	,,
NC-829W;	,,	(# 2023-04)	,,
NC-831W;	,,	(# 2025-96)	,,
NC-833W;	,,	(# 2026-86)	,,
NC-838W;	,,	(# 2027-08)	,,
NC-839W;	,,	(# 2028-88)	,,
NC-871W;	,,	(# 2029-90)	,,
NC-872W;	,,	(# 2030-09)	,,

Registration numbers unknown for serial # 2001 - 2009 - 2024; Bird model A serial #1001 - 1002 - 1009 - 1016- 1018 - 1021 - 1025 - 1048 - 1080 - 1094 - 1095 also eligible for this certificate when modified to conform to specification.

Fig. 131. Waco model CSO with 225 H.P. Wright J6 engine; was typical "straight-wing" with added refinements.

The Waco biplane model CSO was more or less a progressive development of the basic "Waco Ten" design and was quite similar to the previous model ASO (10-W) except for improvement in landing gear design, some improvements in the wing framework, and other minor details that were not conspicuous but provided a general up-grading to this fine utility aircraft design. First known as the CS-225 and often just called the "J6 Straight-Wing", the model CSO was powered with the 7 cyl. Wright "Whirlwind" J6 series engine of 225 h.p. Performance was perhaps a shade better than that of the earlier J5-powered ASO, and it shared all the other fine attributes and pleasant character that made the Whirlwind-Waco such a great favorite. Repeating its performance of a year previous, Waco Aircraft entered two of their new model CSO in the grueling 5000 mile National Air Tour for 1929, and they finished one-two. Hard-flying Johnnie Livingston was the acclaimed winner, and veteran Art Davis was a very close second; the competing field they had to beat was an assembly of some of the finest aircraft in the country.

The model CSO as shown, was a 3 place open cockpit biplane of the general-purpose type and was quite typical to the other models of the basic Waco straight-wing configuration. This craft sported a new out-rigger type landing gear which was a bit more durable than the older long-leg type; the CSO was also available as a seaplane on Edo M twin--float gear. Though not built in any great number because of the general slump in aircraft sales, several of the CSO were still in active operation some 10 years later. The Waco model CSO as powered with the 225 h.p. Wright J6 (R-760) engine, was approved for a type certificate number on 9-30-29 and some 17 or more examples of this model were manufactured by the Waco Aircraft Co. at Troy, Ohio. Clayton J. Bruckner was president and general manager; Lee N. Brutus was V.P.; Robert E. Lees was sales manager; Russel F. Hardy was chief engineer; and Freddie Lund was chief pilot.

Listed below are specifications and performance data for the Waco model CSO as powered with the 225 h.p. Wright J6 engine;

Fig. 132. CSO performed well as seaplane.

length overall 22'6"; hite overall 9'2"; wing span upper 30'7"; wing span lower 29'5"; wing chord both 62.5"; wing area upper 155 sq.ft.; wing area lower 133 sq.ft.; total wing area 288 sq.ft.; airfoil "Aeromarine 2A" modified; wt. empty 1628; useful load 972 lbs.; payload with 63 gal. fuel was 365 lbs.; gross wt. 2600 lbs.; max. speed 128; cruising speed 108; landing speed 45; climb 1100 ft. first min. at sea level; ceiling 19,000 ft.; gas cap. 63 gal.; oil cap. 8 gal.; cruising range at 12 gal. per hour was 540 miles; price at the factory field was $7335. Landplane version of the CSO was also eligible with 22 gal. extra fuel, the following figures will apply; wt. empty 1662; useful load 938; payload with 85 gal. fuel was 200 lbs.; cruising range 750 miles; all other figures were typical. The following figures are for seaplane version; wt. empty 1845; useful load 978; payload with 63 gal. fuel was 371 lbs.; gross wt. 2823 lbs.; max. speed 120; cruising speed 100; landing speed 52; climb 1000 ft. first min. at sea level; ceiling 17,500 ft.; seaplane eligible only with 63 gal. fuel and 8 gal. oil.

The fuselage framework was built up of welded chrome-moly steel tubing, faired to shape with wooden fairing strips and fabric covered. Front cockpit had large door for ease of entry and baggage compartment with locking cover was to the rear of the pilot; baggage allowance for the landplane was 60 lbs., and 30 lbs. for seaplane which included anchor, heave line, and tool kit. The wing framework was built up of heavy sectioned solid spruce spar beams, with spruce and plywood truss-type wing ribs; the leading edges were now covered to preserve the airfoil form and the completed framework was covered with fabric. For extra fuel, beside the 63 gal. in the fuselage tank, a gravity-feed tank of 22 gal. capacity was mounted in the center-section panel of the upper wing. The out-rigger landing gear of 78 inch tread used "Aerol" shock absorbing struts; wheels were 30x5 and Bendix brakes were standard equipment. The CSO could be operated on wheels, skis, or floats; float-gear fittings were an integral part of the fuselage structure. The fabric covered tail-group was built up of welded steel tubing; the fin was ground adjustable and the horizontal stabilizer was adjustable in flight. Wiring for navigation lights, dual controls, and a Hartzell wooden propeller were standard equipment; an engine starter, metal propeller, and Townend speed-ring cowl, were available as optional equipment. Though slated to be replaced by the new model CSO, the J5-powered model ASO was still in demand and built on order up to late 1930; the last of the CSO were also built into late 1930. By this time, models such as the Waco F series were in full production. The next Waco development following the CSO was the "Taper Wing" model CTO as discussed in the chapter for ATC # 257 in this volume. The next "straight-wing" development was the Packard Diesel powered HSO as discussed in the chapter for ATC # 333.

Listed below are Waco model CSO entries that were gleaned from various registration records; this listing may not be complete but it does show the bulk of this model that were built:

NC-8509; Model CSO (# A-57) Wright J6-7-225

NC-21M;	„	(# A-151)	„
NC-518M;	„	(# 3002)	„
NC-262M;	„	(# 3003)	„
NC-516M;	„	(# AT-3005)	„
NC-517M;	„	(# AT-3006)	„
NC-604N;	„	(# A-3107)	„
NC-602N;	„	(# 3109)	„
NC-616N;	„	(# C-3118)	„
NC-612N;	„	(# C-3119)	„
NC-656N;	„	(# 3136)	„
NC-634N;	„	(# 3138)	„
NC-673N;	„	(# 3139)	„
NC-671N;	„	(# 3140)	„
NC-674N;	„	(# 3216)	„
NC-675N;	„	(# 3217)	„
NC-696N;	„	(# 3218)	„

Serial # A-57 first as 10-T (ATO) with Wright J5, converted to CSO; serial # A-151 later converted to CTO taper-wing; serial # 3218 on Edo floats; at least one CSO delivered to Shanghai, China.

Fig. 133. Moreland M-1 with Wright J5 engine was high performance monoplane for business or sport.

Several airplanes were designed and built during this period for the traveling salesman or the business-man who was also a sports-man-pilot; an individual who had a job to do and did it much better in an airplane, especially a small high performance airplane that made every trip a lark and gratifying experience instead of just a means of travel from one place to another. Usually this type of owner-pilot took great pride in his airplane and was most always more than just the average type of pilot; it was for this type of buyer that the Moreland M-1 was designed. A good number of hi-performance open cockpit biplanes were leveled at buyers of this type, but only a very few open cockpit monoplanes dared to enter this exclusive field; the Moreland M-1 of proud personality was one of the select few.

The Moreland M-1 as shown, was a parasol-type monoplane that was arranged in a semi-cabin form to offer protection to occupants of the front seat and yet give the pilot, sitting in the rear, the visibility and other advantages of the open cockpit. Powered with the 9 cyl. Wright J5 engine of 220 h.p., the Moreland M-1 delivered a satisfying performance with flight characteristics and maneuverability designed to placate the playful mood, yet stable enough to afford almost complete relaxation for the pilot on longer jaunts; jaunts that could be stretched to nearly 600 miles if need be. Added to these features was the bonus of a roomy cockpit for all occupants, the ability to carry a substantial payload, and a structure planned to withstand abuse and offer ease of maintenance in the field. The M-1 was outfitted with a desk-panel in the dash-board of the front cockpit that folded out into a handy writing desk; a feature slanted to appeal to the busy business-man who did not pilot, and would have occasion to take care of some last-minute work or make notes, while on the way to his appointed destination.

Of the number that were built, practically all were owned by business-men at one time and this limited-use was somewhat of a misfortune for the M-1 because of the serious business-slump across the nation during this particular time; the market for this type of airplane soon became non-existent. Of the few M-1 that were built, most were yet flying many years later; Maj. Livingston Irving, veteran California pilot, foresaw the advantages of payload and utility, so he outfitted an M-1 "parasol" for crop-dusting in the San Francisco valley areas. In 1936, one M-1 was reported "dusting" bananas in Central

America. The Moreland model M-1 as powered with the Wright J5 engine, received its type certificate number on 9-28-29 and some 4 or more examples of this model were manufactured by the Moreland Aircraft, Inc. of El Segundo, Calif. in a plant-site that later became the nucleus for a division of the Douglas Aircraft Co. G. E. Moreland was the founder and president; Fred W. Herman, formerly an aeronautical engineer with the Air Corps, was V.P. in charge of engineering; and J. J. Atwood, formerly of the Army's Wright Field test-center, was also on the engineering staff.

Listed below are specifications and performance data for the Moreland model M-1 as powered with the 220 h.p. Wright J5 engine; length overall 28'6"; hite overall 8'10"; wing span 39'0"; wing chord 84"; total wing area 260 sq.ft.; airfoil "Goettingen" 398; wt. empty 1780; useful load 1020; payload with 60 gal. fuel was 450 lbs.; gross wt. 2800 lbs.; as outfitted for extensive cross-country work, the M-1 could be equipped for extra fuel load, with complete set of navigational and engine instruments for blind-flying, and extra baggage allowance; in this configuration the M-1 weighed 2000 lbs. empty and useful load was cut down to 800 lbs.; max. speed 130; cruising speed 110; landing speed 50; climb 1050 ft. first min. at sea level; ceiling 17,800 ft.; gas cap. normal 60 gal.; oil cap. 5 gal.; normal range at 11 gal. per hour was 550 miles; price at the factory field was $10,750.

The fuselage framework was built up of welded chrome-moly steel tubing of over-size section and heavily faired to shape with wooden fairing strips and fabric covered; zippered inspection areas were provided to service or adjust important components. The front cockpit was sheltered by a large safety-glass windshield with provisions to extend cabin area clear back to front edge of rear cockpit, thus completely enclosing front seat occupants. A convenient step and large door on left side permitted easy entry to front seat and a step also allowed easy entry to rear cockpit; baggage was carried in a compartment behind the pilot's seat and accessible through a large door on the left side. The semi-cantilever wing framework was built up of spruce spar beams and spruce truss-type wing ribs, with careful attention to every detail; the completed framework was covered with cedar plywood panels. External wing bracing was N-type struts of heavy-sectioned streamlined chrome-moly steel tubing; the wide-tread split-axle landing gear was built into a strong truss with the wing brace struts. Oil-draulic struts were used to absorb the shocks and Bendix toe-operated wheel brakes were standard equipment. Wheels were 30x5, tail-skid was of the spring-leaf type, with tail-wheel available as optional equipment. The fabric covered tail group was built up of welded chrome-moly steel tubing; the fin was ground adjustable and the horizontal stabilizer was adjustable in flight. A metal propeller, inertia-type engine starter, navigation lights, fire extinguisher, adjustable pilot seat, fabrikoid upholstery, and complete set of 12 instruments were standard equipment. Despite loss of sales due to business-slump, Moreland continued some development on the basic M-1 design; a later example mounted the 7 cyl. Wright J6 engine of 225 h.p. An experimental parasol-type trainer monoplane with Kinner K5 engine was also tested, but Moreland finally had to give up some time in 1933.

Listed below are Moreland M-1 entries that were gleaned from registration records:

X-273E; Model M-1 (# 101) Wright J5
;   „   (# 102)   „
NC-805M;   „   (# 103)   „
NC-865N;   „   (# 104)   „

X-882K; Model S-3 (# 201) Wright J6-7-225.

Serial # 101 also operated as 1 place on NR-273E; registration number for serial # 102 unknown; serial # 104 sometimes listed as NC-865M; Model S-3 was to have been new series; NC-805M was dusting bananas in Central America during 1936.

FAIRCHILD "FOURSOME", MODEL 42

Fig. 134. Fairchild "Foursome" model 42 with 300 H.P. Wright J6 engine; scaled down version of model 71.

The Fairchild "Foursome" Model 42 was a high-wing cabin monoplane with seating for 4 that was designed to meet the anticipated demand for a fast and comfortable airplane of medium capacity; a size that would be appealing to men of business, flying sportsmen, and those flying service operators who specialized in fast non-scheduled air-taxi operations. With an eye on the practical side, the roomy interior was upholstered in tough-wearing leather, with ample room for a large amount of baggage, and provisions available for a small lavatory and wash-room; the wings could also be folded easily to require minimum room for storage. Built rugged in good Fairchild tradition, the model 42 had performance to spare and flight characteristics were gentle and predictable. A business-house in Oregon found their "42" of great value to them in calling on customers in difficult country; a famous flyer used his "42" for long pleasure jaunts, and other examples of the "42" were serving well in various parts of the country. Had it not been for the business depression in this country at this particular time, it is quite likely that many more of the Model 42 would have been built to render the service intended.

The "Foursome" design was introduced late in 1928 as the Model 41, shown here; powered with the Wright "Whirlwind" J5 engine of 220 h.p., the model "41" incorporated several innovations and many improvements in general arrangement that can be seen in the illustration. Although performance was very satisfactory and utility promised to be a paramount factor, the model 41 was modified into the model 41-A, as shown, which was later redesignated as the model 42. The Model 42 was now quite similiar in lines to the Fairchild FC series, which had been so familiar to many the country over for several years. The model 42 "Foursome" was a high-wing cabin monoplane, seating four in a spacious interior, with more than ample comfort, and power was boosted over the prototype design to offer the high performance that would be expected in a craft of this type. The powerplant was now the new 9 cyl. Wright J6 series engine of 300 h.p. which was ample power reserve for a sprightly performance, allowing operation under the most exacting conditions in weather, terrain, and ground facilities. Whether it was cheaper or more practical to redesign the prototype airplane to more familiar lines and practice of general arrangement, we cannot say, but the Model 42 was much bigger and heavier than the prototype design, and was more or less a scaled-down version of the earlier Model 71; as a consequence, the 300 h.p. was required to deliver comparable

Fig. 135. Fairchild 41 with Wright J5 engine was forerunner to model 42.

performance. The first 3 examples of the Model 42 were certificated 9-10-29 on a Group 2 approval numbered 2-127; this approval was later cancelled out by ATC # 242. The type certificate number for the model 42 as powered with the 300 h.p. Wright J6 engine, was issued 9-30-29 and some 8 or more examples of this model were manufactured by the Fairchild Airplane Mfg. Corp. at Farmingdale, L.I., New York. Sherman M. Fairchild the founder, was now chairman of the board; Ralph C. Lockwood was chief engineer.

Listed below are specifications and performance data for the Fairchild "Foursome" Model 42 as powered with the 300 h.p. Wright J6 engine; length overall 30'6"; hite overall 9'2"; wing span 45'6"; wing chord 72"; total wing area 290 sq.ft.; airfoil "Goettingen" 387; wt. empty 2676; useful load 1570; payload with 100 gal. fuel was 740 lbs.; gross wt. 4246 lbs.; revised wts. for a later version were as follows; wt. empty 2852; useful load 1448; payload 620 lbs.; gross wt. 4300 lbs.; performance was comparable in either case; max. speed 130; cruising speed 109; landing speed 56; climb 720 ft. first min. at sea level; climb in 10 min. was 5350 ft.; ceiling 15,300 ft.; gas cap. 100 gal.; oil cap. 8 gal.; cruising range at 15 gal. per hour was 700 miles; price at the factory field was $12,900.

The fuselage framework was built up of welded chrome-moly steel tubing that was reinforced with chrome-moly gussets at all highly stressed points; the framework was faired to shape with wooden fairing strips and fabric covered. The interior was upholstered in leather, trimmed in grained walnut, and the cabin walls were lined with heavy blankets of Kapoc for sound-proofing & insulation; provisions were available for

heating and ventilation. Front seats were of the individual bucket type and the rear seat was of the bench type that could be folded down for access to a 35 cu. ft. baggage compartment with an allowance for 200 lbs. Provisions were available to convert this baggage compartment into a small lavatory and washroom. All windows were of shatter-proof glass, and entry into the cabin was available through a large door on each side; dual controls and a full complement of engine and navigational instruments were provided. The semi-cantilever wing framework, in two halves, was built up of spruce box-type spar beams, with wing ribs built up of spruce diagonals and plywood gussets; the leading edge was covered in dural sheet and the completed framework was covered in fabric. Ailerons were built up of welded chrome-moly steel tubing and covered in fabric; fuel tanks were mounted in the root end of each wing half. The wings folded easily on a pivot at rear wing spar and lower end of vee-type wing bracing strut. The landing gear was of the normal split-axle type using oleo-spring shock absorbing struts; wheel tread was 96 inches, wheels were 32x6, and Bendix brakes were standard equipment. Pontoons or skis were quickly interchangeable with the wheeled landing gear. The fabric covered tail group was built up of welded chrome-moly steel tubing; the fin was ground adjustable and the horizontal stabilizer was adjustable in flight. A metal propeller, inertia-type engine starter, full swivel tail wheel, and navigation lights, were standard equipment. The next development in the Fairchild monoplane series was the Model 71-A as discussed in the chapter for ATC # 289 in this volume.

Listed below are Fairchild Model 42 entries that were gleaned from registration records:

NC-390;	Model 41	(# 1)	Wright J5-220.
NC-146H;	Model 42	(# 2)	Wright J6-9-300.
NC-81M;	„	(# 3)	„
NC-106M;	„	(# 4)	„
NC-246V;	„	(# 5)	„
NC-757Y;	„	(# 6)	„
;	„	(# 7)	„
NC-361N;	„	(# 8)	„

Serial # 2 first as Model 41-A, later redesignated Model 42; serial # 3 later as 5 place on Group 2 approval numbered 2-203; registration number for serial # 7 unknown.

Fig. 136. Cessna model DC-6A with 300 H.P. Wright J6 engine; top speed 160 with ease.

This new series of Cessna cantilever-winged monoplanes were larger, bulkier, and packed a good bit more power than the slender and compact A-series built previous. Though a departure that was dictated by normal progress in the private airplane field, the configuration was still pretty much as laid out by Clyde V. Cessna a few years back; a design that stood the test of time and taste, without much change, for at least 25 years. Designed to fit the expected need for a high-performance craft such as this, the DC-6A was powered with a 9 cyl. Wright J6 engine (R-975) of 300 h.p.; a combination which translated into a performance that was hard to top in a craft of this type. With a top speed of better than 160 m.p.h., and a cruise and climb to match, the thundering DC-6A became the choice of many sportsman-pilots. Though possessed of brilliant performance, as the specs will bear out, this craft was rather mild-mannered in its behavior and not too difficult to master. Stanley Boynton, an 18 year old sportsman-pilot, flew a DC-6A from Rockland, Me. to Los Angeles in just under 24 hours, and from Los Angeles back to Rockland in 20 hours and 29 minutes as a record for "junior pilots". A DC-6A was flown by Steve Lacey to 14th place in the National Air

Tour for 1929; penalized slightly in the formula used by its excessive power. In races for cabin monoplanes in the 1000 cu. in. class, the DC-6A was virtually unbeaten. Production of the capricious Warner-powered "AW" continued until mid-summer of 1929; by that time the new enlarged plant was ready to operate, so the AW was rather reluctantly phased out and production of the new DC-6 series was begun. The type certificate number for the model DC-6A as powered with the 300 h.p. Wright J6 engine, was issued on 9-30-29; not particularly blessed with orders because of circumstances that developed later in the year, only 24 examples of the DC-6A were manufactured by the Cessna Aircraft Co. of Wichita, Kansas.

Listed below are specifications and performance data for the Cessna model DC-6A as powered with the 300 h.p. Wright J6 engine; length overall 28'2"; hite overall 7'8"; wing span 41'0"; wing chord (mean) 78"; total wing area 268 sq.ft.; airfoil "Cessna" (modified M-12); wt. empty 1932; useful load 1248; payload with 66 gal. fuel was 606 lbs.; gross wt. 3180 lbs.; a gross wt. of 3350 lbs. was later allowed; max. speed 161; cruising speed 130 plus; landing speed 54; climb 1300 ft. first

Fig. 137. Clean lines of DC-6A offered high performance.

min. at sea level; ceiling 18,500 ft.; gas cap. 66 gal.; oil cap. 9.5 gal.; cruising range at 13 gal. per hour was 600 miles; price at the factory field was $11,500., and later lowered to $11,000.

The fuselage framework was built up of heavy sectioned welded chrome-moly steel tubing that was faired to shape with plywood formers and wooden fairing strips; the completed framework was covered in fabric. The cabin interior was plushly lined in mohair cloth and an 18 cu. ft. baggage compartment with allowance for 271 lbs. was placed in back of the rear seat; front and rear entry doors were conveniently placed, and dual controls were pro-

vided. The cantilever wing framework was built up of spruce box-type spar beams with spruce and plywood wing ribs; the leading edge was covered with plywood to the front spar and the complete framework was covered in fabric. Built up as a continuous framework, the cantilever wing was mounted directly to the top longerons of the fuselage; ailerons were of the balanced-hinge type and the two fuel tanks were mounted in the wing flanking the fuselage. Oddly enough, the oil tank for the engine was installed aft of the cabin. The main landing gear struts were oil-draulic legs of 102 in. tread; wheels were 30x5 and Bendix brakes were standard equipment. The engine was well muffled by a nose-type collector ring; a hand-crank inertia-type engine starter, and ground adjustable metal propeller, were also standard equipment. The fabric covered tail-group was built up of welded chrome-moly steel tubing; the fin was ground adjustable and the horizontal stabilizer was adjustable in flight. With the aid of individual wheel brakes and a 360 deg. full-swivel tailwheel, ground maneuvering of this ship was quite easy. Another development in this new series of cantilever monoplanes was the model DC-6B that is discussed in the chapter for ATC # 244 in this volume.

Listed below are Cessna DC-6A entries that were gleaned from registration records:

Fig. 138. Large volume collector ring on Wright J6 was effective silencer.

NC-9864;	Model DC-6A	(# 199)	Wright J6-300
NC-9869;	,,	(# 204)	,,
NC-627K;	,,	(# 206)	,,
NC-635K;	,,	(# 207)	,,
NC-636K;	,,	(# 208)	,,
NC-638K;	,,	(# 209)	,,
NC-655K;	,,	(# 225)	,,

NC-6441;	,,	(# 226)	,,
NC-640K;	,,	(# 227)	,,
NC-651K;	,,	(# 228)	,,
NC-652K;	,,	(# 229)	,,
NC-653K;	,,	(# 230)	,,
NC-6449;	,,	(# 231)	,,
NC-654K;	,,	(# 232)	,,
NC-301M;	,,	(# 233)	,,
NC-647K;	,,	(# 235)	,,
NC-300M;	,,	(# 237)	,,
NC-302M;	,,	(# 238)	,,
NC-306M;	,,	(# 241)	,,
NC-137V;	,,	(# 245)	,,

Fig. 139. Interior of DC-6A looking forward; mohair was popular fabric.

U.S.A.A.F. procured 4 of the model DC-6A in 1942 as the UC-77 for personnel transport.

Fig. 140. View shows sturdy landing gear and convenient doors.

Fig. 141. Cessna DC-6B with 7 cyl. Wright J6 engine of 225 H.P.

A companion version to the new Cessna DC-6A just discussed, was the more economical model DC-6B. The DC-6B was typical in all respects except for the engine installation which was the 7 cyl. Wright J6 of 225 h.p.; though somewhat less flamboyant, with much less verve than the higher-powered DC-6A, the DC-6B was nevertheless a brilliant performer and a top contender in its class. Comfortable for four with a bonus in performance, the DC-6B was ideal and well suited as a family-plane and a good investment for the business-man. Payload capacity of this version was substantial, and several were placed into duty as air-freighters; at least one was converted to haul newspapers to outlying communities to avoid time-loss in delivering up-to-the-minute news coverage. To promote interest in the new design, a DC-6B was flown by Stanley Stanton to 11th place in the 1929 National Air Tour; Mae Haizlip and Elinor Smith, capable aviatrices of this day, both tried out these new Cessna models in the National Air Races and had nothing but praise for their performance and the way they handled. More favorable circumstances surely would have seen this model built in much greater number. A type certificate number for the DC-6B as powered with the 225 h.p.

Wright J6 (R-760) engine was issued on 9-29-29; a total of 25 examples of this model were built by the Cessna Aircraft Co. at Wichita, Kansas.

Primed and ready for volume production of the new series, Cessna was soon faced with the well-known panic that started with the stock-market crash during October of 1929. The economic weakening that followed, destroyed practically all firm markets for aircraft; the Cessna Company now found itself with a new factory (an enterprise that had grown from a one-room shop in a barn, to a 55,000 sq.ft. factory of 5 buildings), tremendous debts, and hardly any sales for its products. The rosy bubble had burst and Cessna had to admit they were in a bad situation. Redoubling efforts, occasional customers were still found for the DC-6A and the DC-6B, and Clyde Cessna's optimism of the future turned efforts to glider production. Glider clubs were formed, over 300 of the Cessna gliders were sold at $398. each, and a complete break-down of the plant was staved off for the time being at least; thus the company managed to stagger through 1930. Eldon Cessna surmised from the activity of others in the aircraft manufacturing business that perhaps the trend would now be towards low-powered craft, craft that would be cheap to buy and cheap to operate; young Eldon designed the FC-1, then the EC-1 and EC-2, but nobody seemed to be in the market for airplanes of any type, and the new models never reached production stages. The factory had now been closed down and the future looked black and quite uncertain; heart-broken, Clyde Cessna had even offered to work full-time without pay, just to keep the plant open in hopes that this thing would soon blow over, but he was locked out and had to walk away from the plant that was his no more. Clyde Cessna's activities during 1931-32-33 were restricted to designing and building airplanes on special order; these custom-made ships were mostly racing aircraft which have left an indelible mark in the annals of aircraft racing history. Clyde V. Cessna the "old master", retired from aviation activities in 1936 at the age of 56 and returned to his farm in Rago, Kansas; his spirit was not yet broken but he was saddened by the turn of events. The Cessna Aircraft Co. never did go bankrupt as so many others did, but it did not produce any aircraft during 1931-32-33. In 1933 a quiet, lanky youngster from the plains of Kansas, came to Cessna Aircraft Co. as an engineer and test-pilot; this fellow Dwane Wallace figured strongly in the reviving of the dormant company and designed the "Airmaster" series which put the company back in

the running again.

Listed below are specifications and performance data for the Cessna model DC-6B as powered with the 225 h.p. Wright J6 engine; length overall 28'2"; hite overall 7'8"; wing span 41'0"; wing chord (mean) 78"; total wing area 268 sq.ft.; airfoil "Cessna" (modified M-12); wt. empty 1871; useful load 1229; payload with 66 gal. fuel was 607 lbs.; gross wt. 3100 lbs.; max. speed 148; cruising speed 125; landing speed 52; climb 900 ft. first min. at sea level; ceiling 17,500 ft.; gas cap. 66 gal.; oil cap. 7 gal.; range at 11 gal. per hour was 685 miles; price at the factory field was $10,000., lowered to $9750., and then lowered to $9500. in March of 1930.

The construction details, appointments, and other arrangements were more or less similar to those discussed in the preceding chapter. The baggage compartment of 18 cu. ft. capacity had allowance for 100 lbs. 30x5 Bendix wheels and brakes, a metal propeller, and inertia-type engine starter were standard equipment.

Listed below are DC-6B entries that were gleaned from various records:

NC-9863; Model DC-6B (# 198)

Wright J6-7-225

NC-9865;   „   (# 200)   „

Fig. 142. Earlier Cessna CW-6 was basis for DC-6 series design.

Fig. 143. Interior of DC-6B looking aft;
baggage behind rear seat.

NC-9866;	„	(# 201)	„
;	„	(# 202)	„
NC-9868;	„	(# 203)	„
;	„	(# 205)	„
NC-9870;	„	(# 210)	„
NC-631K;	„	(# 211)	„
NC-630K;	„	(# 212)	„
NC-634K;	„	(# 213)	„
NC-632K;	„	(# 214)	„

NC-639K;	„	(# 215)	„
NC-644K;	„	(# 216)	„
NC-629K;	„	(# 217)	„
NC-628K;	„	(# 218)	„
NC-633K;	„	(# 219)	„
NC-641K;	„	(# 220)	„
NC-643K;	„	(# 221)	„
NC-642K;	„	(# 222)	„
C-6444;	„	(# 223)	„
NC-303M;	„	(# 240)	„
NC-305M;	„	(# 244)	„
X-145V;	„	(# 246)	„
NC-569Y;	„	(# 275)	„

Registration number unknown for serial # 202 - 205; serial # 224 - 234 - 236 - 239 - 242 - 247 are in question, they could have been either models DC-6A or DC-6B; Group 2 approval numbered 2-221 issued for DC-6B airplanes convertible from 4PCLM to 1 place freighter, applicable to serial # 200 - 201 - 202 - 211 - 213 - 216 - 217 - 218 - 219 - 222; DC-6B was U.S.A.A.F. model UC-77A that were procured as used aircraft in 1942 for personnel transport.

Fig. 144. Cessna primary glider; Eldon Cessna at controls.

A.T.C. #245
(9-30-29)
BELLANCA "PACEMAKER FREIGHTER", PM-300

Fig. 145. Bellanca "Pacemaker" model PM-300 freighter with 300 H.P. Wright J6 engine.

Bellanca continued on with their tried and true formula for speed with maximum efficiency in the design of their latest "Pacemaker" version. This new craft was called the "Freighter" model PM-300 and was more or less a modified CH-300 design with seating for 4 and available space in the rear section of the cabin for up to 850 lbs. of cargo. Knowing that several of the model CH and CH-300 were most often carrying mixed loads of passengers and cargo, which always made necessary the removal of some seats, Bellanca arranged the PM-300 version with 4 seats in front, a separate baggage compartment for passengers and an enclosed freight compartment of 34 cubic foot capacity; due to anticipated usage of this craft in the so-called "bush country", interior coachwork was arranged more practical and a lot less plush. For bulky loads beyond the capacity of the freight compartment, 3 of the passenger seats could yet be removed quickly to create more floor area; with only a pilot on board, the payload allowance for the PM-300 was a rousing 1575 lbs. This was very unusual indeed for a single engined airplane of this power and size. It is then well apparent that the performance and ability of the "Freighter" version was sure to uphold Bellanca's world-wide reputation for utility and speed with utmost efficiency.

The "Pacemaker Freighter" model PM-300 was a high wing cabin monoplane of typical Bellanca design and was powered with a 9 cyl. Wright J6 series engine of 300 h.p. It is worthy of comment that aerodynamic efficiency allowed this craft to carry a larger useful load than its actual empty weight; an accomplishment that was possible by only a very few airplanes of this period. To further add to its utility, the PM-300 could operate either on wheels, skis, or seaplane float-gear. As a shake-down test for this new version, the veteran George Haldeman flew a model PM-300 in the grueling National Air Tour for 1929, and finished in 5th place amongst a very determined field of contenders. Haldeman flew this same version in the National Air Tour for 1930, and again finished in 5th place with an average of 139 m.p.h. for the distance, demonstrating a consistency in performance at least. During the National Air Races for 1929 held at Cleveland, O., J. Wesley Smith flew a CH-300 to first place in the Philadelphia-to-Cleveland Derby (class F); in the Cleveland-to-Buffalo Efficiency Race, Haldeman placed first with a "Pacemaker" believed to be the PM-300, and J. Wesley Smith was second with a CH-300. Haldeman was first in the Detroit News Trophy race for speed and efficiency, this making a third time win of this trophy for

Fig. 146. PM-300 was flown by Geo. Haldeman to 5th place in 1930 National Air Tour.

Bellanca; Wes Smith placed 3rd with a CH-300 in a 60 mile race for cabin airplanes, bested only by 2 Wasp-Vegas. During the National Air Races for 1930 held at Chicago, the results were almost identical.

One of the new PM-300 versions was delivered to a manufacturer of oil pumping machinery; they soon found the cabin arrangement well suited for flying mixed loads of personnel and oil pumping machinery to far-away sites in Canadian territory. This dual-purpose layout was also suitable for service on feeder-lines that carried both passengers and cargo from out-lying areas to connect with the main transcontinental system. It is odd that the Bellanca "Freighter" version was seldom specifically mentioned and only a few of these craft were actually registered as PM-300 type, but it is logical to believe that proper modifications would convert any of the existing CH-300 "Pacemaker" type into the "Freighter" version . The type certificate number for the "Freighter" version of the standard CH-300 "Pacemaker" as powered with the Wright J6-9 engine of 300 h.p., was issued on 9-30-29; it is hard to determine just how many examples of this model were built because only about 3 are actually listed as PM-300 series in civil aircraft registration records. We have no record of those that were operating in our neighbouring countries. The PM-300 was manufactured by the Bellanca Aircraft Corp. at New Castle, Delaware. Giuseppe Mario Bellanca was the president and chief engineer; Andrew Bellanca was V.P. and secretary; A. D. Chandler was sales mngr.; and R. B. C.

Noorduyn, formerly with Fokker Aircraft, was V.P. in charge of engineering. Bellanca retained several company pilots for testing and development with George W. Haldeman and J. Wesley Smith doing most of the promotional work.

Listed below are specifications and performance data for the Bellanca model PM-300 "Pacemaker Freighter" as powered with the 300 h.p. Wright J6 engine; length overall 27'10"; hite overall 8'4"; wing span 46'4"; wing chord 79"; total wing area 273 sq.ft.; airfoil "Bellanca" (modified 15% R.A.F.); wt. empty 2290; useful load 2310; payload with 86 gal. fuel was 1575 lbs.; gross wt. 4600 lbs.; max. speed 140; cruising speed 120; landing speed 55; climb 900 ft. first min. at sea level; ceiling 17,000 ft.; normal gas cap. 86 gal.; oil cap. 7 gal.; gas capacity increase to 108 or 112 gal. later allowed with corresponding decrease in payload; normal range at 15 gal. per hour was 650 miles; price at the factory field was $15,050.

The fuselage framework was built up of welded chrome-moly steel tubing, faired to shape with wooden fairing strips and fabric covered. The cabin walls were sound-proofed and insulated with blankets of Balsam Wool, and the cabin interior was upholstered and panelled for rough usage; the enclosed freight compartment had a capacity of 34 cu. ft. for a normal allowance of 650 lbs. when 3 passengers were carried. There was a large sky-light in the roof of the cabin for overhead vision and a baggage compartment in the rear section of the cabin, a compartment of 12 cu. ft. capacity for an allowance of 211 lbs. The front seats had folding backs and 3 of the 4 seats could be quickly removed to create floor area for bulky loads; there were provisions for heating and ventilation, and forward windows were of shatter-proof glass. The wing framework, in 2 halves, was built up of solid spruce spar beams that were routed to an I-beam section, with truss-type wing ribs of spruce and basswood diagonals reinforced with plywood gussets; the leading edge was covered with plywood to the front spar and the completed framework was covered in fabric. The unusual "lift struts" were a composite structure of steel tubes and wood formers covered in fabric; of 18 inch chord, the struts were actually airfoils that added 47 sq.ft. to the lifting area and helped to improve the stability. Fuel tanks were mounted in the root end of each wing half; fuel tanks of 43-54-56 gal. capacity were available and all were quickly installed or removed from the top side.

For emergency transport into or out of difficult country, the Bellanca "Pacemaker" could be disassembled into easy-to-handle components; the fuselage was in two halves with a joint just aft of the wing, the wing panels were in two halves and wing struts, landing gear and tail-group, could be removed in small sections for easy transport. The landing gear of 90 in. tread was of the split

axle type and used oleo-spring shock absorbing struts; wheels were 32x6 and Bendix brakes were standard equipment. The fabric covered tail-group was a composite structure of welded steel tubing and wooden formers; the fin was ground adjustable and the horizontal stabilizer was adjustable in flight. A hand-crank inertia-type engine starter, metal propeller, dual stick controls, engine cover, and navigation lights, were standard equipment. A tail-wheel was available as optional equipment, and a seaplane version of the PM-300 was available with Edo K twin-float gear. The next Bellanca development was the 420 h.p. "Skyrocket" which will be discussed in the chapter for ATC # 319, and the next "Pacemaker" development was the model 300-W which will be discussed in the chapter for ATC # 328.

Listed below are Bellanca "Freighter" PM-300 entries that were gleaned from registration records:

NC-257M;	PM-300	(# 159-9)	Wright J6-9-300
NS-2Y;	"	(# 159)	"
;	"	(# 162)	"
NC-860N;	"	(# 183)	"

Serial # 159-9 and # 159 may have been same airplane; serial # 159 was with Dept. of Commerce, Airways Div.; registration number unknown for serial # 162, believed to have been NC-870M; serial # 183 with Aerial Explorations Co.

Fig. 147. Bellanca freighter version ideal in bush country.

Fig. 148. Ford "Tri-Motor" model 7-AT; flown to 3rd place by Myron Zeller in National Air Tour for 1929.

The Ford model 7-AT, though seemingly typical at first glance, was a rare and unusual example in the familiar tri-motor series; unusual at least in the fact that this particular arrangement of airframe and power was primarily developed to participate in the National Air Tour for 1929. Calculated to perform closely to the formula being used for point scoring on the "tour", the 7-AT was flown to 3rd place by Myron E. Zeller in the 5,000 mile reliability contest. This same airplane was also a contestant in the National Air Tour for 1930; tousle-headed Harry L. Russell flew hard to win first place ahead of two very determined "Waco" biplanes. Thus the model 7-AT earned a small measure of fame at least, as the first "Ford" airplane to win the reliability contest, and also as the first 'plane to oust the "Waco" biplanes from their winning ways.

The Ford tri-motor model 7-AT was converted from a model 6-AT (ser. # 6-AT-2) by the installation of a 9 cyl. Pratt & Whitney "Wasp" engine of 450 h.p. in the nose section; the two Wright J6 engines of 300 h.p., in the wing nacelles, were retained for a total power output of 1050 h.p. Arranged as a standard 15 place transport, 13 seats in the main cabin could be quickly removed to provide clear floor area for hauling cargo. Serial # 6-AT-2 the basis for this conversion, was first flown in May of 1929 and converted some months later into a model 7-AT (serial # 7-AT-1). Some time after the winning of the 1930 National Air Tour, this craft was rebuilt at the Ford airplane factory to standard model 5-AT specifications with three "Wasp" engines and test flown in July of 1931. Later in that year it was delivered to American Airways as a 5-AT-C (5-AT-79A). The model 7-AT version was kept on the books as a standard production model, available on order, but only the one example had been built. The type certificate number for the model 7-AT, as powered with one 450 h.p. "Wasp" engine and two 300 h.p. Wright J6 engines, was issued 9-30-29. Only one example of this model was manufactured by the Stout Metal Airplane Co., a division of the Ford Motor Co. at Dearborn, Michigan.

Listed below are specifications and performance data for the Ford-Stout tri-motor model 7-AT-A; length overall 50'3"; hite overall 12'8"; wing span 77'10"; wing chord at root 154″; wing chord at tip 92″; total wing area 835 sq.ft.; airfoil "Goettingen" 386; wt. empty 7280; useful load 5630; payload with 281 gal. fuel was 3400 lbs.; gross wt. 12,910 lbs.; max speed 134; cruising speed 112; landing speed 63; climb 880 ft. first min. at sea level; climb in 10 min. was 6800 ft.; ceiling 14,100 ft.; gas cap. 281 gal.; oil cap. 26 gal.; cruising range at 55 gal. per hour was 560 miles; price at the factory in late 1929 was $51,000., reduced to $47,000. in late 1930; as originally arranged, the 7-AT had no fuel tank in the center-section wing panel, so fuel capacity was less; the following wts. were applicable to this early version; wt. empty 7230; useful load 5330; payload with 235 gal. fuel was 3410 lbs.; gross wt. 12,560 lbs.; performance was slightly better in proportion.

The semi-monocoque fuselage framework was built up of dural channel sections that were riveted together; the outer covering was a .015 inch thick "Alclad" metal skin that was corrugated for added strength and stiffness Passenger chairs were wicker and were quickly removable to provide cargo space of 57x228x67 inch dimensions; a separate baggage compartment had capacity for 30 cu. ft. The full cantilever wing framework was built up of 3 main spar beams and 5 auxiliary spars that were tied together by wing rib trusses; spars and wing ribs were riveted duralumin sections and the completed framework was covered with corrugated "Alclad" metal skin. The landing gear legs, fastened under each outer nacelle, had a wheel tread of 223 inches; wheels were 36x8, covered with huge streamlined wheel fairings, and brakes were standard equipment. The tail-group was built up in a construction similar to the wing

Fig. 149. Tests on Ford tri-motor to eliminate hanging nacelles; conversion not successful.

framework and was also covered with a corrugated "Alclad" metal skin; the horizontal stabilizer was adjustable in flight. Dual "Dep" wheel controls, wheel brakes, tail-wheel, metal propellers, navigation lights, and hand-crank Eclipse inertia-type engine starters, were standard equipment. For a resume of all Ford tri-motor models built previous to this date, we suggest reference to the chapters for ATC # 87 in U.S. CIVIL AIRCRAFT Vol. 1, and chapters for ATC # 132 - 156 - 165 - 173 in Vol. 2. The next development in the Ford tri-motor series was the float-equipped model 5-AT-CS as discussed in the chapter for ATC # 296 in this volume.

Listed below is the Ford 7-AT entry in all its various conversions as gleaned from various records:

-8485; Model 6-AT-A (# 6-AT-2)
    3 Wright J6-9-300.
NC-8485; Model 7-AT-A (# 7-AT-1)
    1 Wasp & 2 J6-9-300.
NC-8485; Model 5-AT-C (# 5-AT-79A)
    3 P & W Wasp.

NC-8485 (as 5-AT-C) operated in American Airways service for several years and was exported to China in 1936.

Fig. 150. Ford 7-AT flown by Harry Russell to first place in 1930 National Air Tour.

DOYLE "ORIOLE", O-2

Fig. 151. Doyle "Oriole" model O-2 with Le Blond 60 engine.

The Doyle "Oriole" model 0-2 was a sporty little parasol monoplane of straight-forward lines that seated two in tandem in open cockpits and was powered with the 5 cyl. LeBlond 60 (5D) of 65 h.p. Designed more or less for sport flying by the private-owner, the configuration was arranged to permit good visibility and the simple and robust structure of this little craft allowed comparative freedom from worry during the average pilot's playful attempt at aerobatics, which was actually flavored more with enthusiasm than precision and was called "stunt flying" back in these days. Performance of this little craft was a very good average for the type, and though referred to as somewhat tricky and capricious by some, it certainly presented no great problems when rigged-up properly or handled with a certain amount of mastery. One outstanding record that stands as verification of the performance built into this little ship, is the flight of Lt. Knud von Clauson-Kaas in a LeBlond powered "Oriole" from Baltimore, Md. to Mexico City. The tiny ship had to fly over mountain peaks with a gross load of 1490 lbs. (120 lbs. overload) and the Mexicans were fairly astounded to see him top 12,000 ft. peaks to reach their capitol; this had been the smallest airplane ever to reach Mexico City whose airport is nearly 8,000 ft. above sea level.

The Doyle Aero Corp. was organized by W. Harvey Doyle and Wilson K. Doyle in the spring of 1928 at Baltimore, Md. and the first example of their new design was quite appropriately called the "Oriole". It was rolled out for flight tests by Otto Melamet about mid-October of 1928; certification tests were conducted into 1929 and a Group 2 approval numbered 2-88 was issued 7-17-29. A type certificate number for the 0-2 as powered with the 65 h.p. LeBlond engine, was issued on 10-2-29 which superseded the Group 2 approval issued previously. The perky "Oriole" was designed by W. Harvey Doyle and Raymond Crowley, young men out of the University of Detroit only a few years previous. Prior to formation of the Doyle Aero Corp., the Doyle brothers had teamed with Jan Pavelka and Dwight Huntington to help form the Vulcan Aircraft Corp. which built the Vulcan "American Moth" parasol monoplane, a design that sometime later became the Davis V-3; because of the death of Wm. Burke, principal financial backer, and very little stock buying interest in the company, the Doyle brothers sold out and left eastward for another start. During the balance of 1929 and into 1930, possibly some 14 examples of the "Oriole" had been built and production capacity had reached to one complete airplane per week; W. Harvey Doyle was the president;

Fig. 152. Trim "Oriole" was ideal for sport flying.

Allan Davis was V.P.; and C. L. Primrose was the secretary. Later in 1930 the crippling business depression had virtually eliminated all sales for small private airplanes, a gutting fire had severely damaged the factory and the Doyle Aero Corp. became bankrupt and forced to call it quits.

Listed below are specifications and performance data for the Doyle "Oriole" model 0-2 as powered with the 65 h.p. LeBlond engine; length overall 19'0"; hite overall 7'6"; wing span 30'0"; wing chord 66"; total wing area 165 sq.ft.; airfoil "Clark Y"; wt. empty 792; useful load 529; payload with 24 gal. fuel was 204 lbs.; gross wt. 1321 lbs.; max. speed 100; cruising speed 85; landing speed 40; climb 680 ft. first min. at sea level; ceiling 16,000 ft.; gas cap. 24 gal.; oil cap. 2 gal.; range at 5 gal. per hour was 350 miles; price at the factory was $2995.

The fuselage framework was built up of welded steel tubing and the wing attach cabane was welded to the longerons as a permanent part of the structure. Duralumin fairing panels were attached to the top and bottom of the fuselage from the firewall back to the tail-post to carry out the oval shape; a door was provided on the left side for ease of entry to the front cockpit, and a step was provided as an entry aid to the rear cockpit. The completed fuselage framework was covered in fabric. The wing framework in two halves, was built up of solid spruce spar beams and spruce truss-type wing ribs; a fuel

Fig. 153. View shows first two "Oriole;" prototype in background, Otto Melamet (test pilot) in foreground.

Fig. 154. "Oriole" with 7 cyl. Leblond engine of 90 H.P.

tank was mounted in the root end of each wing half and the completed framework was covered in fabric. Wing bracing struts were large diameter chrome-moly steel tubes that were encased in balsa wood fairings of a streamlined shape. The landing gear was a chrome-moly steel tube structure of typical 3-member vees that were faired to a stream-lined shape and a spool of shock-cord was attached to the lower end of the vees; wheels were 24x4 and no brakes were provided. The fabric covered tail-group was built up of welded chrome-moly steel tubing and sheet steel formers; the horizontal stabilizer was adjustable in flight. Early versions of the 0-2 were rigged flat with no dihedral in the wing,

but certificate specifications called for 1 deg. of positive dihedral to improve handling characteristics. All "Orioles" were powered with the 5 cyl. LeBlond (5D) engine, but one was tested with the 7 cyl. LeBlond 90 and the last ship was modified in a version powered with the 125 h.p. inverted 4 cyl. air-cooled Martin 333 that was originally designed by Louis Chevrolet, the internationally famous racing-car builder and driver.

Listed below are "Oriole" 0-2 entries that were gleaned from various records:

-10047;	Oriole 0-2	(# A-1)	LeBlond 60
NC-483;	„	(# A-2)	„
-9361;	„	(# A-6)	„
NC-9423;	„	(# A-8)	„
NC-981H;	„	(# A-9)	„
NC-810M;	„	(# A-10)	„
-10598;	„	(# A-11)	„
NC-2242;	„	(# A-)	„
-10218;	„	(# A-14)	„

No registration numbers available for serial # A-3, A-4, A-5, A-7, A-12, A-13; whether they were actually built or not, we cannot say; serial # A-14 later as 0-2 special with Martin 333 engine; serial number unknown for NC-2242; serial # A-11 later had LeBlond (7D) 90 h.p. engine.

Fig. 155. Stubby "Oriole" was airplane of unusual lines, commendable ability.

Fig. 156. Ireland "Neptune" N2C with 450 H.P. Wasp engine; ample room for five.

The amphibious aircraft which could operate off both land and water was a sensible approach in the design of an airplane that was required to offer the utility and convenience of picking and choosing its airports; choosing so they be close to destination, whether it be far inland or on the shores of some body of water. The amphibious aircraft actually saved the time and offered the convenience that made air-travel worthwhile. Such a versatile airplane was the Ireland "Neptune" model N2C; a sea-worthy flying boat with a retracting wheeled undercarriage that could be quickly extended for landings on an airport, or pulled up out of the way for landings on water. The "Neptune" model N2C was a biplane arrangement that had its powerplant mounted rather high in "pusher" fashion between the upper and lower wing panels, where the engine and propeller would be far enough away from the water-spray. The engine and propeller in the rear would minimize the danger to those boarding or leaving the cabin, with the engine running. The sturdy boat-type hull contained a cabin in the forward section, which seated up to 5 persons in ample room and comfort; the forward position of the cabin offered panoramic vision to the pilot and passengers. As

shown here on the "Neptune", the "pusher-type" engine installation offered other advantages; it was easier and safer for docking and anchoring, and its rearward location tended to muffle the engine noises enough to allow near-normal conversation in the cabin. There was naturally some sacrifice in performance to be expected in the "amphibian aircraft", because of its somewhat greater weight and increased parasitic drag, a drag largely due to the greater increase in frontal area. It stands to reason that this sacrifice was indeed a small price to pay for all the extra benefits derived from this type of craft.

The 5 place Ireland "Neptune" model N2C was typical to the earlier N2B in most all respects except for the powerplant installation which was now the 9 cyl. Pratt & Whitney "Wasp" engine of 450 h.p. Performance was substantially increased by the added power and the ability to carry a much greater payload under the most exacting conditions, added up to utility that was hard to beat. Big Bear and Arrowhead are two fairy-land lakes that are nestled high in the Sierras of California; many pilots of sporting blood had hoped to find an airplane (practical) that

Fig. 157. Engine mounting on N2C; "pusher" mounting most practical for this type of airplane.

could scale the summit (6750 ft.) of these lofty pine-clad peaks and swoop down to a landing on one of these enchanting lakes. Retold stories relate of one amphibian craft that flew over these peaks and had actually reached the lake; later it attempted a take-off but remained stuck to the water's surface until the unloading of passengers, extra fuel, chairs, and most everything else that was not fastened down. Only then did it manage a wobbly takeoff. Another "amphibian" got there and stayed there, until it was finally disassembled and carried out in small sections. These high altitude lakes had long been a challenge to the sportsman-pilot of this day, but the "Wasp" powered N2C, the latest development in the "Neptune" amphibian series, reached either of these lakes safely and then took off again easily in 20 seconds, with pilot, 3 passengers, and a full load of fuel. The N2C with its extra power reserve offered a high performance from land or water, with quick and easy take-offs, and exceptional climb-out.

The type certificate number for the "Neptune" N2C as powered with the 450 h.p. "Wasp" engine was issued 10-4-29 and some 9 or more examples of this model were manufactured by Ireland Aircraft, Inc. at Garden City, Long Island, New York. Bertram Work was the president; G. Sumner Ireland was V.P. in charge of design and sales development; D. J. Brimm, Jr. was secretary, treasurer, and chief of engineering; and Wm. Ulbrich was

chief pilot. The Curtiss Flying Service handled all sales for the "Neptune" but Ireland organized his own sales group in 1930. G. Sumner Ireland was associated with Curtiss activities, both directly and indirectly, for many years and his first airplane designed for commercial use was the "Comet" of 1925. The "Comet" was basically a Curtiss "Oriole" fuselage with a new set of wings. In 1926, Ireland introduced the all-new "Meteor" which was an open cockpit biplane carrying 3 or 4 on the power of a 90 h.p. Curtiss OX-5 engine. Ireland offered the amphibian as early as 1926; it was an open cockpit flying boat with an Anzani engine. The first "Neptune" was built in 1927 as an open cockpit craft powered with a Wright J5 engine; several of this version were built but performance was somewhat marginal so Ireland had to go to 300 h.p. in the N2B, which was then a very satisfactory airplane. The sweptback upper wing, which was introduced on the N2B, induced stable flight and served as an easy mark of recognition for the "Neptune" amphibian.

Listed below are specifications and performance data for the Ireland "Neptune" model N2C as powered with the 450 h.p. "Wasp" engine; length overall 31'0"; hite overall on wheels 12'3"; wing span upper 40'0"; wing span lower 34'0"; wing chord upper 72"; wing chord lower 66"; wing area upper 222 sq.ft.; wing area lower 154 sq.ft.;

total wing area 376 sq.ft.; airfoil "Curtiss C-72"; wt. empty 3294; useful load 1566; payload with 80 gal. fuel was 856 lbs.; gross wt. 4860 lbs.; max. speed 120; cruising speed 103; landing speed 48; climb 950 ft. first min. at sea level; ceiling 14,000 ft.; gas cap. 80 gal.; oil cap. 8 gal.; range at 24 gal. per hour was 340 miles; price at the factory was $27,500., later reduced to $23,000.

The hull of the N2C was a rugged framework built up of welded chrome-moly steel tubing, that was surrounded with "Alclad" bulkheads and formers to establish the hull shape and provide a base for fastening the "Alclad" outer metal skin. The cabin in the forward portion of the hull seated two in front and a wide bench-type seat in back had ample room for 3. The main fuel tank was in the hull behind the cabin section; a gravity-feed auxiliary tank was in the center-section panel of the upper wing. Tie-down shackle for mooring was in the nose of the hull, the anchor and heave line was stored in a compartment just ahead of the windshield. Baggage allowance was 176 lbs. which included tools, anchor, rope, and bilge pump. The wing panels were built up of routed spruce spar beams that were reinforced with duralumin and plywood webs at points of great stress, with a combination of wood and metal wing ribs; the leading edges were covered with dural sheet and the completed framework was covered in fabric. The interplane struts were of heavy gauge dural tubing in streamlined section, and interplane bracing was of streamlined steel wire; dural framed and dural covered wing-tip floats were provided to keep the lower wing out of water. The retracting landing gear of 94 inch tread could be drawn up in 35 seconds and lowered in 20 seconds; shock absorbers were air-oil struts and wheels were 32x6. The fabric covered tail-group was a composite structure of duralumin and steel; the fin was ground adjustable and the horizontal stabilizer was adjustable in flight. All movable control surfaces were aerodynamically balanced. Wheel brakes, navigation lights, 3-bladed metal propeller, electric inertia-type engine starter, battery and generator, anchor, and heaving line, were standard equipment. The next Ireland "amphibian" development was the 2 place "Privateer" as discussed in the chapter for ATC # 370.

Listed below are Ireland N2C entries that were gleaned from registration records:

NC-107H;	Model N2C	(# 33)	Wasp 450
NC-86K;	,,	(# 41)	,,
NC-90K;	,,	(# 44)	,,
NC-182M;	,,	(# 45)	,,
NC-89K;	,,	(# 46)	,,
NC-183M;	,,	(# 47)	,,
NC-184M;	,,	(# 48)	,,
NC-186M;	,,	(# 50)	,,
NC-187M;	,,	(# 51)	,,

Serial # 33 was first an N2B with Wright J6-9-300; serial # 44 with Gillam Airways in Alaska.

Fig. 158. Versatility of N2C amphibian made flying extra enjoyable; party shown on flight to mountain lake.

A.T.C. #249
(10-4-29)
MONO "MONOSPORT", MODEL 1

Fig. 159. Monosport model 1 with 110 H.P. Warner engine, had performance enough to delight anyone.

Participation in the annual National Air Races, National Air Derbies, and the many other airplane race events held locally about the country, had a tendency to accelerate normal development and bring out new breeds of airplanes. Standard production models, though also raced in stock configuration were often modified by reducing wing area, fitted with extra streamlining to gain a bit more speed, and quite often were packed with a little more horsepower to gain the advantage over other racing craft in a certain category, or to allow entry with a fair chance to win in a higher power category. These modifications to reduce drag and the boost in horsepower changed the shape of things and quite often led to the development of a new line of sport-type craft that were direct descendants from the lower powered standard models. A case in point is the "Monocoupe 113 Special" which was but a modified model "113" that was faired very carefully for more speed and the power was nearly doubled by the installation of a 110 h.p. Warner "Scarab" engine; from this craft came the direct development of the "Monosport" series. In mid-1929, Henry A. "Tony" Little bought one of the first Warner-powered "Monosport" for sport-flying and air-racing. In approximately 3 years time, the little "Monosport" and its deft pilot had competed in 87 closed-course races, finishing first in 51 events and second in 27 events. This "fast-flying duo" were also contenders in 7 cross-country derbies, taking a first in 3 of them and a second in two of them. Such performance and unflinching stamina was not unusual in Mono-built aircraft.

The flashy "Monosport" was a high performance craft especially leveled at the sportsman-pilot with a low budget and a slight bent to participate in air-racing occasionally, or one who would have need to go on short cross-country trips in the minimum of time. Short-field performance of the "Monosport" series was just short of terrific, so one would not be limited to operate in and out of the larger air-fields; a small pasture-airport was entirely sufficient. By way of description, the "Monosport" model 1 was a stubby two place high-wing cabin monoplane seating two side by side in fairly chummy comfort, with ample visibility in most all directions. Though basically similar to the "Monocoupe", the wing for the "Monosport" was redesigned with less area and graceful elliptical wing tips; the split-axle landing gear was of the short-legged type with oleo ("Mono-oil") shock absorbers as also used on the "Monoprep" trainer. A large cabin door and a convenient step was provided for ease of entry and a large sky-light in the cabin roof afforded vision upwards. Well appointed with many extras, the "Monosport" was a true sports-type craft and its behavior and performance under most any conditions or circumstances left very little to be desired. The "Monosport" model 1 as powered with the 7 cyl. Warner "Scarab" engine was first approved on a Group 2 certificate numbered 2-134 (issued 10-4-29) and its type certificate number was received on the same day; some 8 or more examples of this model were manufactured by Mono Aircraft, Inc. at Moline, Ill.

Listed below are specifications and performance data for the "Monosport" model 1 as powered with the 110 h.p. Warner engine; length overall 21'5"; hite overall 7'1"; wing span 32'3"; wing chord 60"; total wing area 133.2 sq.ft.; airfoil "Clark Y"; wt. empty 1056; useful load 594; payload with 32 gal. fuel was 205 lbs.; gross wt. 1650 lbs.; max. speed 129; cruising speed 110; landing speed 45; climb 1000 ft. first min. at sea level; ceiling 18,000 ft.; gas cap. 32 gal.; oil cap. 4 gal.; cruising range at 6.5 gal. per hour was 520 miles; price at the factory field was $6350., lowered to $4500. in May of 1930.

Construction details and general arrangement of the "Monosport" were similar to the "Monocoupe" as described in the chapter for ATC # 113. A metal propeller, wiring for navigation lights, and wheel brakes were standard equipment; a Heywood air-operated engine starter was available as optional equipment. The "Monosport" model 1 was denied its promised career but it did lay the ground-work for the development of the "Monocoupe 110". A companion development in the "Monosport" series was the Kinner-powered model 2 as described in the chapter for ATC # 250 in this volume.

Listed below are "Monosport" model 1 entries as gleaned from registration records:

NC-8957; Monosport 1 (# 2000) Warner 110.
NC-8968;   „   (# 2001)   „
NC-8974;   „   (# 2002)   „
NC-105K;   „   (# 2005)   „
    ;   „   (# 2007)   „
NC-145K;   „   (# 2011)   „
NC-152K;   „   (# 2012)   „
NC-161K;   „   (# 2014)   „
NC-167K;   „   (# 2015)   „

Serial # 2000 - 2001 - 2002 - 2005 were on Group 2 approval numbered 2-134; type certificate number # 249 was for serial # 2007 and up; registration number for serial # 2007 unknown.

Fig. 160. Monosport popular with sportsman-pilots; often used for racing.

Fig. 161. Monosport also used in business promotion.

A.T.C. #250
(10-4-29)
MONO, "MONOSPORT" MODEL 2

Fig. 162. Monosport model 2 with 100 H.P. Kinner K5 engine.

The "Monosport" Model 2 which seemed to present a more saucy appearance than the Model 1 discussed just previous, was a companion model in this series, that was more or less typical except for its powerplant installation; the power mounted in this case was the popular 5 cyl. Kinner K5 engine of 100 h.p. The "Monosport 2" as shown, was also a high-winged cabin monoplane of diminutive proportions that seated two side by side and performance was more or less identical, so interest in either of these two models would be but a matter of engine preference. Characteristics differed between the two ships only to the extent that the Kinner engine was a bit rougher in operation and tended to set up vibrations that were a bit annoying at times. Several of the "Monosport 2" were owned by sportsmen who were pilots that raced their craft in contest on occasion, and one example was operated by the Kinner Airplane & Engine Co. as a test-bed and sales promotion demonstrator. Flown by Leslie H. Bowman, Kinner Engine's test pilot, this craft placed 2nd in Class D category of the Miami to Cleveland Air Derby held in 1929.

Veteran pilot Vernon L. Roberts, who did most of the test-flying and a lot of the development work at Mono Aircraft, was largely instrumental in the developing of the "Monosport" series; other famous "Mono" pilots at this particular time were "Ike" Stewart, R. T. "Stub" Quimby, H. A. "Tony" Little, and of course Phoebe Omlie, the hard-flying Tennessee belle. The "Monosport" series made themselves rather conspicuous in contest and helped lay a firm foundation for other famous "Monocoupe" models that were to follow in the years to come. The "Monosport" Model 2 as powered with the 100 h.p. Kinner K5 engine was first certificated on a Group 2 approval numbered 2-135 for serial # 2003 and # 2004; an approved type certificate number for the series was issued 10-4-29 and in all some 7 or more examples of the "Monosport 2" were manufactured by the Mono Aircraft Corp. at Moline, Illinois. Don A. Luscombe wielded a guiding hand as the president; Clayton Folkerts rustled up a constant flow of genius as the chief engineer; and veteran Vern Roberts passed judgement on all as the chief pilot.

Listed below are specifications and performance data for the "Monosport" Model 2 as powered with the 100 h.p. Kinner K5 engine; length overall 21'5"; hite overall 7'1"; wing span 32'3"; wing chord 60"; total wing area 133.2 sq.ft.; airfoil "Clark Y"; wt. empty 1053; useful load 597; payload with 32 gal. fuel was 208 lbs.; gross wt. 1650 lbs.; max. speed 129; cruising speed 110; landing speed 45; climb 990 ft. first min. at sea level; ceiling 17,500 ft.; gas cap. 32 gal.; oil cap. 4 gal.; cruising range at 6 gal. per hour was 550 miles; price at the factory field was $5750., lowered to $4250. in May of 1930.

The fuselage framework was built up of welded 1025 and chrome-moly steel tubing, faired to shape with steel tube and wooden fairing strips and fabric covered. The cabin interior was tastefully upholstered in leather and fine fabric, and a good-sized baggage shelf was behind the seat back; because of the short-coupled and fairly high engine mounting, visibility from the "Monosport" was restricted directly forward to some extent. There was good visibility to the sides and a large skylight in the cabin roof provided good vision upward, a boon to pilots in a steep turn around a pylon. The robust wing framework was built up of heavy-dimensioned spruce spar beams that were routed out to an I-beam section, and truss-type wing ribs were built up of basswood webs and spruce cap-strips; the leading edge was covered to preserve the airfoil form and the completed framework was covered in fabric. Gravity-feed fuel tanks of 16 gal. capacity were mounted in the root end of each wing half, which were actually spliced together in the center to form a one-piece wing. The short-legged split axle landing gear was built up of chrome-moly steel tubing and used oil-draulic shock absorbing struts; wheels were 28x4 and brakes were standard equipment.

Fig. 163 Monosport 2 also popular with sportsman-pilots; shown here at factory.

The fabric covered tail-group was built up of welded steel tubing; the fin was ground adjustable and the horizontal stabilizer was adjustable in flight. A metal propeller, wiring for navigation lights, and dual stick controls, were also standard equipment; an engine starter of either the inertia-type or air-operated type such as the Heywood, were optional equipment. Construction details as listed above were typical for both models of the "Monosport". The promising career of the "Monosport 2" was cut painfully short because of economic circumstances, but it later evolved into the "Monocoupe 125" which will be discussed in the chapter for ATC # 359. The next "Mono" development to be discussed in this volume, is the "Monocoach 275" as approved on ATC # 275.

Listed below are "Monosport 2" entries that were gleaned from registration records:

NC-8980; Monosport 2 (# 2003) Kinner K5.
NC-8989;   „   (# 2004)   „
NC-113K;   „   (# 2006)   „
NC-136K;   „   (# 2008)   „
NC-142K;   „   (# 2009)   „
NC-144K;   „   (# 2010)   „
 -153K;   „   (# 2013)   „

Serial # 2003 - 2004 on Group 2 approval numbered 2-135; serial # 2006 Kinner Motors test-bed; serial # 2013 also had 7 cyl. LeBlond 90 h.p.

STEARMAN "BUSINESS SPEEDSTER", C3R

Fig. 164. Stearman "Business Speedster" model C3R with 225 H.P. Wright J6 engine;
ideal for business or sport.

Incorporating several refinements in detail and a general up-grading, the "Stearman" model C3R was but a progressive development in the basic C3 series design. Loaded with extras to provide easier operation and fitted with some finery lacking in the earlier C3B, the model C3R turned out to be about 100 lbs. heavier when empty but available payload and general performance was still comparable. To attract more customers in a generally lucrative field, the model C3R was more or less leveled to the needs of business flying and was appropriately called the "Business Speedster". Several examples did fall heir to various chores in the business world but most of the C3R found their ways into hands of the flying services for general-purpose work or into hands of sportsman-pilots who wanted a capable and spirited

airplane that would deliver top-notch performance. Stearman performance and reliability was already a reputation of such stature in the aviation industry that customers could buy airplanes from only a factory description and be reasonably sure of getting what they had bargained for. As a consequence, the prestige of the "Stearman" name was so high in flying circles the country over that when a prospective customer heard that an improved "Stearman" biplane was to be built around the new 225 h.p. Wright J6 engine, he placed his order sight unseen before even a prototype model was built.

The Stearman model C3R was an open cockpit biplane of what was loosely called the general-purpose type, with seating for 3; the upholstered front cockpit was over 33 in.

wide with a volume capacity of over 30 cu. ft. to allow ample room and good protection for two passengers. Of uncomplicated lines originated several years ago, this new "Stearman" biplane was a handsome airplane that also bore signs of unmistakeable class and good breeding. Powered with the improved 7 cyl. Wright J6 (R-760) engine of 225 h.p., the model C3R delivered a profitable utility and very good performance for general-purpose work; the new series Wright engine was also a boon to flying service operators in the way of less maintenance and somewhat greater reliability for more hours on the flight-line. Sure-footed and extremely capable, the C3R likewise shared the spirited flight characteristics and the pleasant behavior that made the "Stearman" biplane such a great favorite. The rugged framework and inherent reliability naturally tended to promote longevity so it is not hard to accept the fact that at least 20 of the model C3R were still in active service some 10 years or so later. First approved on 10-7-29 as a land plane with 2700 lbs. gross, the type certificate number for the model C3R as powered with the 225 h.p. Wright J6 engine was amended 10-27-29 for landplane with increased gross weight of 2754 lbs. This certificate was later amended to include the seaplane version of the C3R on Edo P floats with a 3050 lbs. gross weight. A tally shows that some 29 or more examples of the model C3R were manufactured by the Stearman Aircraft Co. at Wichita, Kansas; a division of the United Aircraft & Transport Corp. The all-metal "Hamilton" monoplane (now a Boeing subsidiary) was slated to be moved from the plant in Milwaukee, Wis. to a plant-site alongside of Stearman Aircraft, but contemplated production never did materialize. Lloyd C. Stearman was the president and general manager; Mac Short was V.P. and the chief of engineering; Hall Hibbard was assistant engineer; Walter Innes, Jr. was the secretary; and J. E. Schaefer was sales manager.

Listed below are specifications and performance data for the "Stearman" model C3R as powered with the 225 h.p. Wright J6 engine; length overall 24'11"; hite overall 9'0"; wing span upper 35'0"; wing span lower 28'0"; wing chord upper 66"; wing chord lower 54"; wing area upper 186 sq.ft.; wing area lower 108 sq.ft.; total wing area 294 sq.ft.; airfoil "Stearman" # 2 or # 6; wt. empty 1741; useful load 1013; payload with 65 gal. fuel was 403 lbs.; gross wt. 2754 lbs.; max. speed 130; cruising speed 110; landing speed 47; climb 1000 ft. first min. at sea level; ceiling 17,500 ft.; gas cap. 65 gal.; oil cap. 7 gal.; cruising range at 12.5 gal. per hour was 550 miles; price at the factory field was $8500. The following figures are for seaplane version of C3R on Edo P twin-float gear; wt. empty 2037; useful load 1013; payload 400 lbs.; gross wt. 3050 lbs.; max. speed 120; cruising speed 100; landing speed 55; climb 880 ft. first min. at sea level ceiling 15,500 ft.; cruising range 500 miles. The following figures are for prototype airplane as approved 10-7-29; wt. empty 1741; useful load 959; payload with 65 gal. fuel was 349; gross wt. 2700 lbs.; all other specifications and performance figures were more or less comparable.

Fig. 165. C3R shared typical Stearman lines of classic beauty.

Fig. 166. Rugged C3R was well at ease on small pasture-airports.

The fuselage framework was built up of welded chrome-moly steel tubing, faired to shape with wooden fairing strips and fabric covered. The fuselage was faired out to a slightly larger cross-section, metal fairing panels were added to the forward fuselage sides back to the lower wing, the lower wing was streamlined into the fuselage by a small metal fillet, and a baggage compartment of 10 cu. ft. capacity with 60 lb. allowance for land-plane and 30 lb. allowance for seaplane was located in back of the rear cockpit. The front cockpit was well upholstered and had a door on left side for entry; the pilot's cockpit was well placed, had a streamlined head-rest, and well placed foot-steps for ease of entry. The wing framework was built up of solid spruce spars that were routed to an I-beam section with spruce and plywood truss-type wing ribs; the leading edges were covered with dural sheet and the completed framework was covered in fabric. The main fuel tank was mounted high in the fuselage just ahead of the front cockpit, an additional fuel tank was mounted in the center-section panel of the upper wing. Ailerons were in the upper wing only and were actuated by two push-pull rods that came out of the pilot's cockpit and into the center-section panel; here they were connected to bell-cranks and push-pull tubes that provided differential action. The split

axle landing gear of 90 inch tread was of the out-rigger type and used oleo-rubber shock absorbing struts; wheels were Kelsey-Hayes or Bendix 30x5 and brakes were standard equipment. The fabric covered tail-group was built up of welded chrome-moly steel tubing; the fin was ground adjustable and the horizontal stabilizer was adjustable in flight. A metal propeller, and wiring for navigation lights was provided as standard equipment. A hand-crank inertia-type engine starter, cockpit and navigation lights, and custom finishes, were offered as optional equipment. The next development in the "Stearman" biplane was the "Wasp" powered model 4-E as described in the chapter for ATC # 292 in this volume.

Listed below are Stearman model C3R entries that were gleaned from registration records:

NC-8828;	Model C3R	(# 5001)	Wright J6-225
NC-8822;	„	(# 5002)	„
NC-8840;	„	(# 5003)	„
NC-659K;	„	(# 5004)	„
NC-657K;	„	(# 5005)	„
NC-658K;	„	(# 5006)	„
NC-656K;	„	(# 5007)	„
NC-660K;	„	(# 5008)	„
NC-661K;	„	(# 5009)	„
NC-662K;	„	(# 5010)	„

Fig. 167. C3R was well-behaved seaplane.

NC-668K;	,,	(# 5011)	,,	NC-790H;	,,	(# 5034)	,,
NC-669K;	,,	(# 5012)	,,	NC-793H;	,,	(# 5035)	,,
NC-670K;	,,	(# 5013)	,,	NC-794H;	,,	(# 5036)	,,
NC-671K;	,,	(# 5014)	,,	NC-799H;	,,	(# 5037)	,,
NC-672K;	,,	(# 5015)	,,	NC-566Y;	,,	(# 5038)	,,
NC-673K;	,,	(# 5016)	,,	R-567Y;	,,	(# 5039)	,,
NC-675K;	,,	(# 5017)	,,
NC-773H;	,,	(# 5018)	,,
NC-775H;	,,	(# 5019)	,,
NC-780H;	,,	(# 5030)	,,
NC-781H;	,,	(# 5031)	,,
NC-782H;	,,	(# 5032)	,,
NC-789H;	,,	(# 5033)	,,

Serial # 5001 was originally listed as serial # 249 before C3R was given # 5000 serial block; there is no listing for serial # 5020 thru # 5029 and no explanation available for this gap in continuity which involves 10 airplanes; serial # 5039 converted as C3P with Wright J5 on Group 2 approval 2-445.

Fig. 168. C3R shown was last of 38 aircraft built in this series.

Fig. 169. Lockheed "Vega" 2-A was rare 7 place version with 300 H.P. Wright J6 engine.

Most basic aircraft designs that were built to any number, were bound to show up occasionally in modified or special-purpose versions. Adaptations such as these were powered or arranged to suit a calculated demand, or perhaps the demands of a particular customer. The versatile Lockheed "Vega" was particularly noted for this straying from the normal because the basic design was quite flexible and was so easily adapted into a combination suited to a specific need. Among the several variants stemming from the standard "Wasp-Vega" (the Wasp-Vega was by now the standard version since use of the 220 h.p. Wright J5 engine was more or less discontinued, and use of the 300 h.p. Wright J6 was fairly limited), this was the one and only "Vega" model 2-A that was built as a 7 place airplane with a 300 h.p. power-plant. Arranged in maximum seating with the economical operation of a smaller engine, the model 2-A was no doubt planned as an air-taxi for executive use; the P. Lorrillard Co., manufacturers of "Old Gold" cigarettes, were

registered as the original owner, but record of its use and disposition are not conclusive enough to verify this. Despite the uncertainty of its ownership and operation, we do know that the "Vega" 2-A was the only one of its kind and such a "rare bird" that even a photo of this ship was impossible to get.

The somewhat mysterious Lockheed "Vega" model 2-A was a high-winged cabin monoplane seating 7 and it was powered with the 9 cyl. Wright J6 series (R-975) engine of 300 h.p. The cabin area seated six passengers and the pilot was more or less isolated up in "the front office", which was elevated slightly above the cabin level; the pilot's windshield formed a cabin enclosure that was faired into the leading edge of the cantilever wing. Undoubtedly typical in configuration, it is safe to assume that flight characteristics or general behavior would not have been altered too much in this combination but performance suffered and the verve the normal "Vega" was noted for, surely was not present in this version. Specifications and general lay-out reveal that it was designed as an economical air-taxi to haul the maximum in payload with a minimum of power; that it was not entirely satisfactory in this arrangement is hinted at by reports that it was slated for export to a foreign country and also that it was to be dismantled. The possibility that it was later rebuilt to mount the 450 h.p. "Wasp" engine has also been rumored. The type certificate number for the Lockheed "Vega" model 2-A was issued 10-7-29 and only one example of this model was manufactured by the Lockheed Aircraft Corp. at Burbank, Calif.

Sometime in July of 1929 the assets of the Lockheed Aircraft Corp. were acquired by the Detroit Aircraft Corp., a combine of Detroit business-men that had plans of setting up a "General Motors" type organization of the air industry. With control already over Ryan, Parks, Eastman, and several other small companies, the organization had fond hopes of a great success but soon found that Lockheed was the only one in their operation bringing in any measure of profit. To oversee the Lockheed plant, Detroit Aircraft Corp. officers, headed by Edward S. Evans, sent down Carl B. Squier as general manager; Squier, an experienced salesman and pilot from Michigan, was a competent man of friendly personality that turned out to be a shot-in-the-arm to Lockheed personnel whose morale had sagged due to the change in ownership. Presiding over the engineering department was Gerard "Gerry" Vultee who had several great plans for the future and two assistants; his two gifted draftsmen-engineers were Jimmy Gerschler and Richard Von Hake. Dependable and competent "Herb" Fahy was the company test-pilot and assisting him on occasion was Wiley Post, who had not yet become famous for his round-the-world flights.

Listed below are specifications and per-

formance data for the Lockheed "Vega" model 2-A as powered with the 300 h.p. Wright J6 engine; length overall 27'6"; hite overall 8'6"; wing span 41'0"; wing chord at root 102"; wing chord at tip 63"; wing chord mean 80"; total wing area 275 sq.ft.; airfoil at root Clark Y-18; airfoil at tip Clark Y-9.5; wt. empty 2305; useful load 1915; payload with 100 gal. fuel was 1085 lbs.; gross wt. 4220 lbs.; max. speed 155; cruising speed 133; landing speed 55; climb 900 ft. first min. at sea level; ceiling 16,500 ft.; gas cap. max. 100 gal.; oil cap. 8-10 gal.; cruising range at 15 gal. per hour was 800 miles; price at the factory field was $14,985.

Construction details and general arrangement for the "Vega" 2-A were typical to other models as described in the chapters for ATC # 49 - 93 - 140 - 169 - 227. The cigar-shaped fuselage was a wooden monocoque construction with seating for 7, with a baggage compartment to the rear of the cabin section with allowance for 95 lbs. The one-piece full cantilever wing was built up of box-type spar beams and Pratt truss-type wing ribs; the completed framework was covered with 3/32 inch plywood. The long-leg landing gear of 96 inch tread used 30x5 wheels and individual brakes. A metal propeller and hand-crank inertia-type engine starter were standard equipment. The next Lockheed development was the low-winged "Sirius" model 8 as described in the chapter for ATC # 300 in this volume.

Listed below is the only known entry for the Lockheed "Vega" 2-A as gleaned from registration records:

NC-505K; Vega 2-A (# 83) Wright J6-9-300.

Serial # 83 was manufactured 8-29, approved 10-7-29; Lockheed memo dated 9-24-30 states ship was dismantled but this does not seem likely; may have been exported; no record of use or disposition.

Fig. 170. Model TP-W was improved Swallow trainer with 110 H.P. Warner engine.

The "Swallow" TP-W version was a companion model to the OX-5 powered TP and the Kinner powered TP-K; the model TP-W was typical in most all respects except for its powerplant installation which was the 7 cyl. Warner "Scarab" engine of 110 h.p. Rumor first had it that the TP-W would be modified somewhat with a split-axle landing gear and other deluxe finery to better qualify it also as a sport-plane for the private-owner flyer, but these plans must have been discarded and it emerged as a typical TP trainer except for the change in powerplants. Performance was slightly increased with the added power, but its relatively high price tag no doubt discouraged sales among the flying-schools who had to watch their budgets very carefully these days; as a consequence, the Swallow TP-W was built only in small number and remained a rather rare craft.

The "Swallow" model TP-W was an open cockpit biplane that seated two in tandem and though typical of the TP series in many respects, it was faired out into somewhat better lines and some changes were made in the windshields, their arrangement, and the cockpit cut-outs. Designed primarily as a training craft able to handle secondary phases of pilot-training, the stiff-legged TP-W was quite well-behaved, rather sprightly in response, and it shared all the other fine attributes that made the Swallow TP series such a great favorite the country over. Of those that received their pilot training in the TP, most surely remember it with a great fondness. The type certificate number for the Swallow TP-W as powered with the 110 h.p. Warner "Scarab" engine was issued 10-9-29 and some 2 or more examples of this model were manufactured by the Swallow Airplane Co. at Wichita, Kansas. W. M. Moore was the president; H. Mitchener was now V.P.; Geo. H. Bassett was secretary and general manager; L. H. Connell was now the chief pilot; and Dan Lake was the chief engineer.

Listed below are specifications and performance data for the "Swallow" model TP-W as powered with the 110 h.p. Warner engine; length overall 23'9"; hite overall 8'10"; wing span upper 30'11"; wing span lower 30'3"; wing chord both 60"; wing area upper 150 sq.ft.; wing area lower 146 sq.ft.; total wing area 296 sq.ft.; airfoil USA-27 modified; wt. empty 1201 lbs.; useful load

538; payload with 28 gal. fuel was 170 lbs.; gross wt. 1739 lbs.; max. speed 102; cruising speed 87; landing speed 37; climb 760 ft. first min. at sea level; ceiling 16,000 ft.; gas cap. 28 gal.; oil cap. 4 gal.; cruising range at 7 gal. per hour was 340 miles; price at the factory was announced at $4600. but soon lowered to $4250.

Construction details and general arrangement as described in the chapters for ATC # 105 and # 186 of U.S. CIVIL AIR-CRAFT, Vol. 2 will also apply for the model TP-W as described here. About 100 of the OX-5 powered Swallow TP had already been built, and several of these were later modified to TP-K specifications with the Kinner K5 engine; the model TP-K was catching on and selling fairly well and the Warner powered model TP-W was now marking the end of the line. As the TP design was slowly phasing out of production, the Swallow Co. turned to the development of the "Model H" series, a 3-seated open sport-craft of pleasant lines and very excellent performance; this move being prompted by a feeling that a good market for this type of airplane was yet available, despite the economic depression which was to stifle

certain other facets of aeronautical activity. For discussion of the "Swallow HA" refer to the chapter for ATC # 341.

Listed below are "Swallow" model TP-W entries that were gleaned from various registration records:

NC-8166;	Model TP-W	(# 116)	Warner 110
NC-8187;	,,	(# 202)	,,
NC-425N;	,,	(# 205)	,,

Serial # 116 was first as model TP; no baggage allowed for model TP-W; all TP-W also eligible with 125 h.p. Warner engine.

Passing a flight exam was most often a heart-breaking event that hinged its success on the almighty "inspector" and often governed by the mood he was in; because of this, many stories have been told throughout the years, but the following is one of the best. After a check flight the student-pilot was emphatically notified that he almost "stalled" on take-off, all his turns were flat, he had crossed the controls on figure eights, leveled off much too high, and made a very bum landing. With that the student quipped; Well, did I climb in and out of the cockpit O.K.?!!

Fig. 171. Travel Air model 4-D with 225 H.P. Wright J6 engine; handsome 4-D was very popular.

The handsome and chesty-looking "Travel Air" model 4-D was a direct development from the Wright J5 powered model B-4000 and was typical in most respects except for its powerplant installation which was now the 7 cyl. Wright J6 engine of 225 h.p. Designed especially for the sportsman-pilot and other limited-use services, the model 4-D had plenty of eye-appeal and enough honest performance to satisfy even the most critical; the 4-D was rugged, had utility and a substantial payload to offer, plus a broad range of good all-round performance. The robust airframe could absorb plenty of hard work or punishment and the eager flight characteristics of this craft often prompted an exhuberance into a pilot that translated itself into aerial gyrations of pure enthusiasm; a display not always precise but at least enthusiastic. Had the market been more favorable for this type of aircraft, it is most likely that many more of the model 4-D would have been built.

Basically the "Travel Air" model 4-D was a 3 place open cockpit biplane of the so-called speed-wing type; the "elephant-ear" ailerons so characteristic of the "Travel Air" had since been discarded and the new wing cellule now had well-rounded wing tips and were beefed-up considerably to handle the extra power, weight, and performance. The earlier "Travel Air" being lighter had a tendency to float a bit when landing but the 4-D had heft that was not easily distracted from its path in. The powerplant for the model 4-D was the 7 cyl. Wright J6 (R-760) series engine of 225 h.p. which was somewhat lighter than the J5 "Whirlwind" and output was increased by an average of 5 h.p.; performance of this latest development in the "4000" series was commendable indeed and it was generally judged to be one of the finest versions of the "Travel Air" biplane line. The type certificate number for the model 4-D as powered with the J6 "Seven" was issued 10-12-29 for both the landplane and seaplane; some 19 or more examples of this model were manufactured by the Travel Air Co. at Wichita, Kansas. Feeling the pinch of the "depression" and anticipating some trouble ahead, the Travel Air Co. was soon to join the huge Curtiss-Wright Corp. as a part of the Curtiss-Robertson division; they felt perhaps there would be safety in numbers to cope with the lean period that lie just ahead.

Listed below are specifications and performance data for the "Travel Air" model 4-D as powered with the 225 h.p. Wright J6 engine; length overall 23′4″; hite overall 9′1″; wing span upper 33′0″; wing span lower 28′10″; wing chord upper 66″; wing chord lower 56″; wing area upper 171 sq.ft.; wing area lower 118 sq.ft.; total wing area 289 sq.ft.; airfoil T.A. # 1; wt. empty 1837; useful load 1043; payload with 67 gal. fuel was 428 lbs.; gross wt. 2880 lbs.; max. speed 130; cruising speed 110; landing speed 52; climb980 ft. first min. at sea level; ceiling 14,000 ft.; gas cap. 67 gal.; oil cap. 6 gal.; cruising range at 14 gal. per hour was 520 miles; price at the factory field was first quoted at $7960., raised to $8640. in May of 1930. The following figures are for the model 4-D as seaplane on Edo P twin-float gear; wt. empty 2044; useful load 956; payload with 67 gal. fuel was 341 lbs.; gross wt. 3000 lbs.; max. speed 122; cruising speed 104; landing speed 58; climb 900 ft.; ceiling 13,000 ft.; gas cap. 67 gal.; oil cap. 6 gal.; range at 14 gal. per hour was 490 miles; price was about $1500. extra for float gear installation.

The fuselage framework was built up of welded chrome-moly steel tubing, faired to shape with wooden fairing strips and covered in fabric. The roomy cockpits (each about 30x38) were well upholstered and there was a good-sized baggage compartment just behind the rear seat with a capacity of 9 cu. ft. and allowance for 50 lbs. The wing framework was built up of solid spruce spar beams that were routed out for lightness, with spruce and plywood truss-type wing ribs; the leading edges were covered to preserve the airfoil form and the completed framework was covered in fabric. The main fuel tank was mounted high in the fuselage just ahead of the front seat, and extra fuel was carried in a tank that was mounted in the center-section panel of the upper wing. The robust landing gear of 84 in. tread was of the out-rigger type and employed oleo-spring shock absorbing struts; wheels were 30x5 and Bendix brakes were standard equipment. A metal propeller, navigation lights, and hand-crank inertia-type engine starter were also standard equipment. A metal cover for front cockpit, and a speed-ring engine cowling were available as optional equipment. Later models of the 4-D mounted the Wright J6 (R-760) engine of 240 h.p. with a slight increase in performance. The next "Travel Air" development was the model 10-D, a 4 place cabin monoplane as discussed in the chapter for ATC # 278 in this volume; the next "Travel Air" biplane development

Fig. 172. Speed-ring cowling and wheel pants boosted speed of 4-D.

was the Jacobs-powered model 4-P as discussed in the chapter for ATC # 280, also in this volume.

Listed below are Travel Air model 4-D entries that were gleaned from registration records:

NC-8137; Model 4-D (# 898) Wright J6-7-225
NC-9916; „ (# 1102) „
NC-9961; „ (# 1160) „
NC-9990; „ (# 1166) „
NC-696H; „ (# 1263) „
NC-692H; „ (# 1264) „
NC-695H; „ (# 1265) „
NC-689K; „ (# 1270) „
NC-697K; „ (# 1320) „
NC-699K; „ (# 1322) „
NC-440N; „ (# 1327) „
NC-442N; „ (# 1329) „
NC-448N; „ (# 1360) „
NC-455N; „ (# 1361) „
NC-472N; „ (# 1362) „
NC-465N; „ (# 1363) „
NC-480N; „ (# 1364) „
NC-153V; „ (# 1366) „
NC-467N; „ (# 1367) „

Serial # 1102 later converted to B9-4000; serial # 1160 also as 4-D Special; serial # 1263 sometimes listed as BE-4000; serial # 1265 was first as BE-4000; serial # 1367 later on Group 2 approval numbered 2-300.

Fig. 173. 4-D blessed with rugged landing gear and beefy wings for care-free sport flying.

Fig. 174. Star "Cavalier" C with Le Blond 60 engine; pert "C" was rare version.

The Star model C was a progressive development in the "Cavalier" series, an interesting small cabin monoplane of normal arrangement and configuration that seated two side by side in average comfort; the model C was powered with the 5 cyl. LeBlond 5-D engine of 65 h.p. as a companion version to the model B. Arranged so as to appeal to the average fixed-base operator or private owner, it also offered good flight characteristics and ample performance for an airplane of this type; the small engine offered good economy and the simple and rugged structure offered many hours of flying with a minimum amount of repairs or maintenance. Although possessing all of the normal features for an airplane of this type, plus a very pleasant nature and good value, the "Cavalier" was in a highly competitive class so sales were spread out quite thin over a long period of time.

The "Cavalier" model C as pictured here, was a high-winged cabin monoplane of

pleasant proportion seating two side-by-side and was more or less typical to the earlier model B except for the powerplant installation, which was now the 65 h.p. LeBlond 5-D engine; the little LeBlond which traced its parentage back to the Detroit "Air Cat" engine, was becoming quite popular in this power range. In performance there was hardly any noticeable difference between the models B or C, and both were as proper as any young school-girl, so selection of either would be but a matter of engine preference. The "Cavalier" monoplane was finally built in about six different models and about 55 examples in all, were reported built over a production period of some 5 years. The type certificate number for the "Cavalier" model C, as powered with the 65 h.p. LeBlond engine, was issued 10-12-29 and some one or more examples of this model were manufactured by the Star Aircraft Co. at Bartlesville, Okla. John H. Kane was president; Wm. D. "Billy" Parker was V.P., general manager, and in charge of sales; E. A. "Gus" Riggs was the chief engineer. The company's directors were also officers of the Phillips Petroleum Co., one of the largest independent oil companies in the U.S.A. "Billy" Parker and "Gus" Riggs who teamed up to design and develop the "Cavalier" monoplane series, were men that drew from experience gained since 1912; both men were noted for advancements they introduced to early aviation.

Listed below are specifications and performance data for the Star "Cavalier" model C as powered with the LeBlond 5-D engine; length overall 19'8"; hite overall 6'3"; wing span 31'6"; wing chord 61"; total wing area 156.85 sq.ft.; airfoil Clark Y; wt. empty 861 lbs.; useful load 539; payload with 25 gal. fuel was 204 lbs.; gross wt. 1400 lbs.; max. speed 100; cruising speed 85; landing speed 37; climb 580 ft. first min. at sea level; ceiling 10,000 ft.; gas cap. 25 gal.; oil cap. 2.5 gal.; cruising range at 4.5 gal. per hour was nearly 450 miles; price at the factory was $2985.

The fuselage framework was built up of welded chrome-moly steel tubing in a rigid Warren truss form; the framework was faired to shape with wooden formers and fairing strips then fabric covered. The cabin interior was trimmed and upholstered neatly, with large cabin windows and an over-head skylight to insure the maximum in visibility; there was a large baggage shelf behind the seat-back with allowance for 39 lbs. The wing framework in two halves, was built up of solid spruce spars that were routed to an I-beam section, with spruce and plywood truss-type wing ribs and the completed framework was covered in fabric; the fuel supply was carried in two tanks that were mounted in the root end of each wing-half, flanking the fuselage. The landing gear was of the normal split-axle type using two spools of rubber shock-cord to absorb the bumps; wheel tread was 61 inches, wheels were 24x4 or 22x10 and no brakes were provided. The fabric covered tail-group was built up of welded chrome-moly steel tubing; the fin was ground adjustable to compensate for high-r.p.m. torque and the horizontal stabilizer was adjustable in flight. The next development in the "Cavalier" series was the model D as powered with the 80 h.p. "Genet" (British) engine and the "Lambert 90" powered model E as discussed in the chapter for ATC # 321.

Listed below are "Cavalier" model C entries as gleaned from registration records: NC-993H; Model C (# 110) LeBlond 60.

It is possible that serial # 118 may have been a model C but it is not listed in registration records.

Fig. 175. Davis model D-1 with Le Blond 60 engine; parasol wing had double taper.

The petite Davis parasol-type monoplane was an enchanting and strictly for-fun airplane that certainly enjoyed a good measure of lasting popularity during its active lifetime. It was neat and trim, quite girlish, and very well suited for the average pilot-owner who wanted to enjoy owning his own airplane and to fully enjoy its friendly nature and eager response to command. It is known fact that enthusiasm and pride among owners always ran high for this airplane, in all the various models; its charm most likely lies in the happy balance of responsive performance, durability of structure, and its pleasant uncluttered lines. The Davis model D-1 was a 2 place open cockpit "parasol" monoplane that seated two in tandem; the front cockpit for the passenger was placed under the center-section panel of the wing, and the pilot's cockpit in the rear, was placed well for entrance and visibility. The stout semi-cantilever wing was rather unusual in that it had double taper from the wing-strut station, tapering to the tip and also to the center-section, both in plan-

form and in section. The rugged wing structure was strut-braced to the fuselage and its elevated position afforded the maximum in visibility, produced a good pendulum stability and the many other desirable features inherent to this high-wing placement. Davis owners just never seemed to get disgruntled about anything, so it is little wonder that quite a few are still flying or being restored, and it has been one of the most sought after light airplanes of this period.

The model D-1 was an improved version of the Davis V-3; it was powered with the 5 cyl. LeBlond 5D engine of 65 h.p. or the 5DE of 70 h.p. and its type certificate number was issued on 11-8-29. Some 9 or more examples of this model were manufactured by the Davis Aircraft Corp. of Richmond, Indiana. Walter C. Davis, president of the firm and former Capt. in the Air Service, was also a fine pilot and thought nothing of ferrying a new Davis monoplane to some distant customer; Lt. Lewis "Pat" Love, another fine

Fig. 176. Wing placement on D-1 offered excellent visibility.

pilot, was the factory manager and the chief pilot; Dwight Huntington, one of the original engineers on the developing of the "American Moth", was retained as the chief engineer.

The "Davis" monoplane had a very interesting early history that starts out with the designing and building of the Vulcan "American Moth" by Harvey Doyle and Jan Pavelka; an extremely interesting design that was engineered to a great extent by Dwight W. Huntington, who has had a hand in several other designs of merit during this period. An "American Moth", modified somewhat and powered with a Warner "Scarab" engine of 110 h.p., made its appearance late in 1928 and left its presence noticeably felt under the deft hands of Robert Dake at numerous air-races and air-derbies around the country. Despite the performance dished-out by the "American Moth", and its potential as a sport-craft, stock buying interest in the Vulcan Aircraft Co. of Portsmouth, Ohio was somewhat lacking so the firm and its assets were finally sold to Walter C. Davis and then re-organized in Jan. of 1929 as the Davis Aircraft Corp.; Davis then modified the design just a bit into the Davis V-3. The model V-3 was awarded a Group 2 approval numbered 2-119 (issued 9-6-29) and 23 of this popular model were built. A few slight modifications, which were mainly in the tail-group and cockpit interior, were performed on the V-3 design and this later became the model D-1; a basic design that was brought out in at least 3 other power-combinations.

Fig. 177. The Vulcan "American Moth" was basis for Davis design.

Listed below are specifications and performance data for the Davis model D-1 as powered with the 65 h.p. LeBlond 5D engine; length overall 19'10"; hite overall 7'3"; wing span 30'2"; wing chord max. 63"; wing chord mean 56"; total wing area 145 sq.ft.; airfoil "Goettingen" 387 (modified) at max. thickness, tapering to a "Clark Y" at root and tip; wt. empty 839; useful load 495; payload with 20 gal. fuel was 190 lbs.; gross wt. 1334 lbs.; max. speed 100; cruising speed 85; landing speed 38; climb 680 ft. first min. at sea level; climb in 10 min. was 5000 ft.; ceiling 10,000 ft.; gas cap. 20 gal.; oil cap. 2 gal.; cruising range at 4.5 gal. per hour was 350 miles; price at the factory was $3285., lowered to $2995. in early 1930, and to $2695 later in the year; the saving was based largely on lower manufacturing costs.

The fuselage framework was built up of welded 1025 and chrome-moly steel tubing, heavily faired to an oval shape and covered in fabric; the front cockpit had an entrance door on the left side, and provisions for 20 lbs. of baggage were made in a compartment behind the rear seat. The wing framework was built up of laminated spruce spar beams that were routed out for lightness, with duralumin wing ribs built up of rounded channel section; the leading edge was covered with dural sheet and the completed framework was covered in fabric. The center section of the wing was built up of welded steel and in it was housed the 20 gal. fuel tank; wing bracing struts were heavy-sectioned steel tubes that were encased in balsa-wood fairings and the center-

section cabane was of streamlined steel tubing. The landing gear was of the split-axle type with two long telescoping legs of 78 in. tread; shock absorbers, at the upper ends, were made up of two spools of rubber shock-cord, and wheels were 26x4. Dual stick controls were provided, with the front stick quickly removable and front rudder pedals could be quickly disconnected. The fabric covered tail-group was built up of welded steel tube spar members, with sheet steel former ribs; the fin was ground adjustable and the horizontal stabilizer was adjustable in flight. The next development in the Davis "parasol" monoplane was the Kinner powered model D-1K; refer to the chapter for ATC # 272 in this volume.

Listed below are Davis D-1 entries that were gleaned from registration records:

Registration	Model	Serial	Engine
NC-733V;	Davis D-1	(# 125)	LeBlond 60
;	,,	(# 126)	,,
NC-854N;	,,	(# 127)	,,
NC-855N;	,,	(# 128)	,,
NC-856N;	,,	(# 129)	,,
NC-641N;	,,	(# 130)	,,
NC-642N;	,,	(# 131)	,,
NC-643N;	,,	(# 132)	,,
NC-644N;	,,	(# 133)	,,

Serial numbers eligible for ATC # 256 were # 125 and up; registration number for serial # 126 unknown; serial # 128 later converted to model D-1-66 with LeBlond 85; all serial numbers listed also eligible with 70 h.p. LeBlond 5DE engine.

Fig. 178. Davis V-3 was earlier model; improved version became D-1.

Fig. 179. High performance and friendly nature made D-1 a long-time favorite.

A.T.C. #257
(10-16-29)
WACO "TAPER WING", CTO

Fig. 180. Waco taper wing model CTO with 225 H.P. Wright J6 engine.

Since 1928 the fabulous Waco "Taper Wing" continued on as one of the most exciting and most colorful airplanes of this period; first flying to fame in late 1928 by winning the transcontinental air derby from New York to Los Angeles, the tapered wing "Waco" had been more or less in the spotlight ever since. At the 1928 National Air Races held that year in Los Angeles, Calif., the new "Taper Wing" finished 2-3-4 in a 75 mile free-for-all; also in that year the first commercial airplane to perform the outside-loop was a Waco "Taper Wing" flown by fearless Freddie Lund. The following year in the Portland, Ore. to Cleveland, O. Air Derby, the "Taper Wings" came in 2-5-6; at the National Air Races for 1929 held that year in Cleveland, Art Davis brought his 10-T (Taper Wing) from the rear of the pack to 2nd place in the Australian pursuit race. In a 60 mile· race for women, Gladys O'Donnell bested all in her new CTO "Taper Wing"; she came in first in the Australian pursuit race for women and next day flew the same race to 3rd place under handicap. In the civilian acrobatic exhibition, the Waco team of 3 "Taper Wings" led by Freddie Lund, took top honors. In the

women's Pacific Derby of 1930 from Long Beach, Calif. to Chicago, Ill., Gladys O'Donnell gunned her CTO "Taper Wing" to first place; taking a first again a few days later in a women's free-for-all, and first in an 800 cu. in. event for open ships. In the men's Air Derby from Miami to Chicago, the "Taper Wings" were 1-2; in a race for open ships of 800 cu. in., Lloyd O'Donnell and Art Davis showed everyone the way by finishing 1-2 in their Waco CTO. To wind up events in the 1930 races, "Taper Wings" placed 1-2-3 in the balloon bursting contests and were tops again in the civilian acrobatic exhibition. Results were more or less the same the next year, and the next; it was actually some time before the flash and the glitter of the Waco "Taper Wing" was eclipsed to any extent. The foregoing account was really not meant to be a coverage of the National Air Races but only to picture briefly the type of service the Waco "Taper Wing" reveled in. Because of its very nature the "Taper Wing" was lifted above any hum-drum duties and spent most of its life frolicking in "stunts" for a multitude of spectators to marvel at, or nudging its flashing wings dangerously close to some closed-course pylon.

The Waco "Taper Wing" model CTO (225-T) was also a 3 place open cockpit biplane with seating for 3 and was quite typical to the previous model ATO except for its engine installation, landing gear arrangement, and a few minor changes necessary to this combination. Powered with the 7 cyl. Wright J6 (R-760) engine of 225 h.p., the new model CTO was also fast, eager, flashy, and highly maneuverable; because of its lighter empty weight there was a considerable increase in its payload capacity. As noted in the preceding paragraph, the CTO brought in its share or better of the winnings and was every bit as good an airplane as the Wright J5-220 powered ATO. Some say not and will quote the fact that the ATO was still built into 1931 and often ordered in preference to the J6-225 powered CTO; methinks that this was but a matter of engine preference — the old J5 really was a well-loved engine. Built in fairly good number and well scattered about the country, the Waco "Taper Wing" had a personality that created an interested gathering of admirers everywhere it went; it is not surprising that a good number were still flying some 10 years later and a few were still flying in the "sixties". The model CTO was the last "Taper Wing" in production but owners were continually seeking better performance and modifying previously built

craft to the extent that some could sustain nearly 180 m.p.h. around the pylons of a racing course. Actually stressed to operate with as much as 450 h.p., there was some talk of plans to redesign the "Taper Wing" into a 2 place high-performance "fighter" for export to foreign countries but plans of this nature never did materialize; with the plans shelved for a time, they were later revived to some extent in the famous Waco "Model D" series. The type certificate number for the Waco "Taper Wing" model CTO, as powered with the 225 h.p. Wright J6 engine, was issued 10-16-29 and some 8 or more examples of this model were manufactured by the Waco Aircraft Co. at Troy, Ohio. Clayton J. Bruckner was president and general manager; Lee N. Brutus was V.P.; R. E. Lees was sales manager; L. E. St. John was secretary; and A. Francis Arcier, formerly of the General Airplanes Corp., succeeded Russell F. Hardy as chief engineer.

Listed below are specifications and performance data for the Waco "Taper Wing" model CTO as powered with the 225 h.p. Wright J6 engine; length overall 22'6"; hite overall 9'0"; wing span upper 30'3"; wing span lower 26'3"; wing chord both at root 62.5"; total wing area 227 sq.ft.; airfoil M-6; wt. empty 1677 lbs.; useful load 923; payload with 65 gal. fuel was 308 lbs.; payload with 100 gal. max. fuel was 98 lbs.; gross wt. 2600

Fig. 181. CTO modified for racing; Tex Rankin in cockpit.

Fig. 182. Taper wing with 300 H.P. Wright J6
engine was JTO.

Fig. 183. Waco "Taper Wing" as fighter plane
for foreign countries.

lbs.; max. speed 137; cruising speed 115; landing speed 52; climb 1200 ft. first min. at sea level; ceiling 19,000 ft.; gas cap. normal 65 gal.; gas cap. max. 100 gal.; oil cap. 8 gal.; cruising range at 12.5 gal. per hour was 575-890 miles; price at the factory field was $8525.

The fuselage framework was built up of welded chrome-moly steel tubing, faired to shape with wooden fairing strips and fabric covered. The front cockpit had a door for easy entrance and a metal panel was available to close this cockpit off when not in use; a small baggage compartment of 40 lb. capacity was in the turtle-back section behind the rear cockpit. The wing framework was built up of heavy sectioned solid spruce spar beams with spruce and plywood truss-type wing ribs that were closely spaced for added strength and to help preserve the true airfoil form; the completed framework was covered with fabric. The stout wings were tapered in plan-form and in section; there were four ailerons that were connected together in pairs by a streamlined push-pull strut. The interplane struts were of heavy gauge chrome-moly steel tubing in a streamlined section and the interplane bracing was of heavy gauge streamlined steel wire to withstand the abnormal stresses that would be imposed on an airplane of this type. Incidently, the whole "Taper Wing" structure was stressed to take up to 450 h.p. so its safety margin with the 225 h.p. Wright J6 was tremendous. The main fuel tank of 65 gal. capacity was mounted high in the fuselage just ahead of the front cockpit

and extra fuel was carried in two 17.5 gal. tanks that were mounted in the center-section panel of the upper wing. The robust landing gear of 78 inch tread was now of the outrigger type and used oil-spring shock absorbing struts; wheels were 30x5 and Bendix brakes were standard equipment. The fabric covered tail-group was built up of welded chrome-moly steel tubing; the fin was ground adjustable and the horizontal stabilizer was adjustable in flight. A metal propeller, wheel brakes, wiring for navigation lights, dual controls, and custom color combinations were standard equipment. Navigation lights, inertia type engine starter, speed-ring engine cowling, and semi-balloon airwheels were optional equipment. The next Waco development was the famous Model F as described in the chapter for ATC # 311.

Listed below are Waco "Taper Wing" model CTO entries as gleaned from various registration records:

NC-9534; Waco CTO (# A-29) Wright J6-7-225
    NC-21M;   „   (# A-151)   „
    NC-515M;   „   (# A-3004)   „
    NC-265M;   „   (# AT-3007)   „
    NC-655N;   „   (# 3137)   „
    NC-657N;   „   (# 3166)   „
    NC-666N;   „   (# 3117)   „
    NC-688N;   „   (# 3207)   „

Serial # A-29 was first as ATO; serial # A-151 and # 3117 later on Group 2 approval 2-378; any Waco serial number prefixed by letter A or C were eligible for modification to CTO.

Fig. 184. Consolidated "Commodore" model 16 with 2 Hornet B-575 engines.

As a forerunner to the era of the large "flying boat", the twin-engined "Commodore" pioneered a service that was planned to link the major cities of Northern America with those of South America. Formed especially to provide this service the New York, Rio, Buenos Aires, Line (N.Y.R.B.A.) launched schedules in Feb. of 1930 over an 8900 mile route. Leaving the bay in Miami weekly on a 7-day schedule, the 8-stage flight arrived in Buenos Aires at least 11 days before the fastest steam-ship. After christening ceremonies by Mrs. Herbert Hoover (wife of U.S. President Hoover) were over, the flag-ship of the "Commodore" fleet, the "Buenos Aires", embarked with its first load in July of 1929 on a trail-blazing flight to initiate the extended route. Early business was quite brisk and most of the passengers both ways were representatives of business-houses on commercial missions. The N.Y.R.B.A. Line was on the threshold of a great career but it just barely got started. Possibly due to financial difficulties, which plagued many lines at this time, directors of the new line were discussing a possible merger with the Pan American Airways System; in a deal consummated in Sept. of 1930, Pan American bought the air-line and 10 of its "Commodore" flying boats at a cost ranging from $97,000. to as much as

$106,000., depending on the condition of the craft when sold. Acquisition of the line now gave Pan Am over 19,000 miles of airways which when viewed on a map, traced a rough figure-8 from lower U.S. thru' most of So. America. In Oct. of 1930 Pan Am purchased the remaining 4 "Commodore" that were part of the order contracted for by the N.Y.R.B.A. Line in 1929 and this brought their "Commodore" fleet to 14. Especially designed for the longer over-water hops, the "Commodores" with Pan American initiated service in 1930 from the island of Jamaica to Panama; the longest non-stop flight over water in the world at this time. Also that same year, the twin-engined "Commodore" opened a direct route from Miami to the Panama Canal Zone. The huge "Commodore" boats served proudly and very well for years on various portions of Pan American's far-flung system and several were still flying in regular scheduled service as late as 1935.

The Consolidated "Commodore" model 16, often described as the "Leviathan" of the air, was a huge high-winged monoplane of the "flying boat" type with seating for 18-22 passengers and a crew of 3; accommodations were very comfortable, unusually spacious, and there was yet ample room for air-mail

Fig. 185. Leaving Miami "Commodore" heads for So. America on 7-stage flight.

and cargo. Powered with two 9 cyl. Pratt & Whitney "Hornet" B (R-1860) engines of 575 h.p. each, the "Commodore" could easily maintain over 100 m.p.h. in cruising speed over long stretches of open water; maximum cruising range was at least 1000 miles. I. M. Laddon, formerly an engineer with the Air Corps' McCook Field Test Center, joined Consolidated Aircraft in 1927 and became the chief of design and engineering for heavy aircraft. In 1928, Laddon and his staff designed the huge XPY-1 "Admiral" as a long-range patrol boat for the U.S. Navy; a civil-version of this craft became the passenger carrying "Commodore" model 16. This related series were the first big airplanes built by Consolidated and the direct forerunners of the famous PBY "Catalina" of World War 2. The prototype of the "Commodore" differed from the later production versions only in that it had an open cockpit up forward for the pilots; this drafty station was later covered over with a streamlined enclosure. Also, the first model 16 was some 1000 lbs. lighter than later versions; the inclusion of all sorts of extra equipment and fancy appointments in following examples raised the empty weight considerably and this cut into the payload

allowance. The model 16 Type 1 was arranged with seating for a maximum of 25, and the Type 2 had cabin arrangements for a maximum seating of 33; the gross weight was the same in all cases. Pan American records clearly state that seating capacity on the "Commodore" was a variable factor depending mainly on the fuel load carried. The type certificate number for the "Commodore" models 16, 16-1, 16-2, was issued 11-20-29 and 14 examples in this series were manufactured by the Consolidated Aircraft Corp. at Buffalo, New York.

Listed below are specifications and performance data for the Consolidated "Commodore" model 16 as powered with two 575 h.p. "Hornet" B engines; length overall 61'8"; hite overall 15'8"; hite in water 13'8"; wing span 100'0"; wing chord 138"; total wing area 1110 sq.ft.; airfoil Goettingen 398; wt. empty 10,550 lbs.; useful load 7050; payload with 650 gal. fuel was 2255 lbs. (10 passengers & 600 lbs. baggage-cargo); payload with 450 gal. fuel was 3455 lbs. (18 passengers & 485 lbs. baggage-cargo); all figures based on crew of 3; gross wt. 17,600 lbs.; max. speed 128; cruising speed 108; landing speed 60; climb 675 ft. first min. at sea level; climb to 10,000 ft. was 25 min.; ceiling 11,250 ft.; gas cap. max. 650 gal.; oil cap. 50 gal.; cruising range at 64 gal. per hour was 1000 miles; price at the factory was $125,000. The "Commodore" was the highest priced commercial airplane at this time. The following figures are for the 33 passenger model 16-2; wt. empty 10,500; useful load 7100; payload variable with fuel load; gross wt. 17,600 lbs.; all other figures were comparable. The following figures are

Fig. 186. XPY-1 "Admiral" was basis for "Commodore" design.

for prototype model 16; wt. empty 9620; useful load 7980; payload with 650 gal. fuel was 3185; gross wt. 17.600 lbs.; performance was comparable.

The hull framework was built up of extruded duralumin members that were riveted together into a structure of bulkheads and longitudinal stringers; the completed frame was covered with aluminum alloy sheet of varying gauges. The main cabin area of 8 ft. x 18 ft. x 5 ft. dimension was arranged with 3 passenger compartments; the two forward compartments normally seated 8 passengers each and the rear compartment normally seated 4. Individual seats were richly upholstered and fitted with life-preserver cushions; half the seats in each compartment faced forward and half the seats faced to the rear. An aisle provided access from one compartment to the other and access to the cabin area was through a walk-in hatchway in the aft portion of the hull. The pilots cabin was in the forward part of the hull, with storage for mooring gear immediately forward; just in back of the pilots cabin was the radio-room, a compact lavatory for the passengers, and a 200 cu. ft. cargo hold with provisions for 60 cu. ft. of baggage space. The huge wing framework comprised a center section and two outer panels that were built up of Warren truss type spar beams riveted together from aluminum alloy extruded angles, with Warren truss-type wing ribs riveted together from extrusions and channel sections of aluminum alloy; the leading edges were covered with dural sheet and the completed framework was covered with fabric. Fuel tanks were mounted in the center portion of the wing and oil tanks were mounted in each engine nacelle. The semi-cantilever wing was braced in a "parasol" fashion by a multitude of struts that were arranged to provide mounts for the engine nacelles and out-rigger floats; all struts were of aluminum alloy in a streamlined section except those that carried greater loads and stresses. In this case, these were of chrome-moly steel tubing in a streamlined section. The fabric covered tail-group was built up of extruded aluminum alloy sections in a manner similar to that of the wing; the horizontal surfaces were placed atop a central stub fin and double rudders were placed on the top surface of the stabilizer to be well clear of the water. Huge ailerons of a construction similar to the wing were of the Friese balanced-hinge type and the dual rudders used aerodynamic balance to lighten control loads. 3-bladed metal propellers, electric inertia-type engine starters, generator, batteries,

Fig. 187. "Commodore" was good "boat," able to make landings in open sea.

navigation lights and landing lights, and mooring gear were standard equipment. The next Consolidated development was the "Fleetster" model 17 as described in the chapter for ATC # 291 in this volume.

Listed below are "Commodore" entries as gleaned from various records:

X-855M; Commodore (# 1) 2 Hornet B.
NC-658M;   „   (# 2)   „
NC-659M;   „   (# 3)   „
NC-660M;   „   (# 4)   „
NC-661M;   „   (# 5)   „
NC-662M;   „   (# 6)   „
NC-663M;   „   (# 7)   „

NC-664M;   „   (# 8)   „
NC-665M;   „   (# 9)   „
NC-666M;   „   (# 10)   „
NC-667M;   „   (# 11)   „
NC-668M;   „   (# 12)   „
NC-669M;   „   (# 13)   „
NC-670M;   „   (# 14)   „

Serial # 1 - 2 - 3 were model 16; serial # 1 later converted to model 16-1; serial # 4 - 5 - 6 - 7 - 8 - 9 - 12 - 13 -14 were model 16-1; serial # 10 - 11 were model 16-2; serial # 1 - 2 - 3 - 4 - 6 - 7 - 8 - 9 -10-11 were purchased by Pan American from N.Y.R.B.A.; serial # 5-12-13-14 purchased by Pan Am from Consolidated Aircraft.

Fig. 188. N.Y.R.B.A. "Commodore" later flew for Pan American system, were forerunners to era of the "clipper ship."

A.T.C. #259
(10-19-29)
INLAND "SPORT", S-300

Fig. 189. Inland "Sport" model S-300 with Le Blond 60 engine.

The pert and handsome Inland "Sport" was an interesting little open cockpit monoplane of the "parasol" type that seated two in sporty comfort and was powered with the popular 5 cyl. LeBlond 60 (5-D) engine of 65 h.p. Neat, trim, and quite appealing, this small craft was designed primarily to provide the practical companionship of side-by-side seating in the sporty atmosphere of an open cockpit. Quite suitable for pilot training, it was often stressed that it would be a good deal easier and much more efficient to teach a student side-by-side than in tandem. Built rugged for the average private-flyer, the "Sport" was also well able to withstand the strains and hard knocks bound to appear in every-day sport flying and student training. Of good aerodynamic proportion, the flight characteristics of this little monoplane were positive and quite friendly; though not blessed with extra power reserve, the performance was quite adequate for normal use. The type certificate number for the Inland "Sport" model S-300, as powered with the 65 h.p. LeBlond engine, was issued 10-19-29 and some 15 or more examples of this model were manufactured by the Inland Aviation Co. at Fairfax Field in Kansas City, Kansas. Arthur Hardgrave was the president; Lt. Wilfred G. Moore was general manager and sales manager; Milton C. Baumann,

formerly assistant engineer to Waverly Stearman while at Butler Aircraft, was the chief engineer; Allen Smith was assistant engineer; "Bert" Thomas conducted most of the test flying, while Wm. A. "Bill" Ong, Wm. "Bill" Green, and Chas. Dailey were sales representatives that did most of the promotion work in the field.

The original prototype airplane that was later developed into the Inland "Sport", was a 1927 design by "Dewey" Bonebrake; this was a light 2-seated "parasol" monoplane of typical configuration that was reported to be powered with a 6 cyl. Anzani (French) engine. This was highly probable because the light-plane designers and builders of this period certainly had not much to choose from in the way of engines, and the "Anzani" was always available and just about as good as any to be had at this particular time. The prototype airplane by Bonebrake was flown by Gene Gabbert (or Gebhart) to the 1928 National Air Races held in Los Angeles; it was planned to show it off and perhaps secure a sponsor for its manufacture. The new "parasol" no doubt created a good bit of interest, though having to vie for attention with many other new designs that were also present at the races. Arthur Hardgrave, who was later impressed with its character and its possibi-

Fig. 190. "Super Sport" with 110 H.P. Warner engine; Wilfred Moore in foreground.

lities, decided to sponsor the new design for manufacture and the Inland Aviation Co. was formed at Kansas City, Mo. for that purpose; a plant site was later acquired on Fairfax Field in Kansas City, Kansas. L. D. Bonebrake, originator of the design that led to the Inland "Sport", was an engineer and test pilot for Unit Motors & Airplane Co. of Kansas City, Mo. in 1924; he redesigned the Bahl "Lark" (a parasol monoplane sometimes refered to as the Tuxhorn "Lark") to improve its handling and flight characteristics. The show of performance put on with this redesigned airplane by Blaine Tuxhorn as the pilot, was just short of remarkable for these early times.

Bent on promotion for their new craft, an Inland "Super Sport" with the 110 h.p. Warner "Scarab" engine was readied for events at the 1929 National Air Races held at Cleveland, O.; a LeBlond powered "Sport" competed also. Both aircraft showed off well and the high caliber of performance inherent in the design was brought to the flying public's attention. In a speed run over a 100 kilometer course, the "Super Sport" averaged 127 m.p.h. and later this same craft set an altitude record for light airplanes by reaching 19,700 feet. By November of 1929 ten "Sports" had already been built, the original Inland "Sport" had already flown over 35,000 miles in demonstration, a factory had been built and tooled-up, and production was well under way; orders began coming in

from all over the country.

Listed below are specifications and performance data for the Inland "Sport" model S-300 as powered with the 65 h.p. LeBlond engine; length overall 19'10"; hite overall 7'6"; wing span 30'0"; wing chord 60"; total wing area 144 sq.ft.; airfoil M-12; wt. empty 768 lbs.; useful load 524; payload with 24 gal. fuel was 200 lbs.; gross wt. 1292 lbs.; max. speed 100; cruising speed 85; landing speed 35; climb 680 ft. first min. at sea level; ceiling 12,000 ft.; gas cap. 24 gal.; oil cap. 2 gal.; cruising range at 4.5 gal. per hour was 425 miles; price at the factory field was $3485. but substantially reduced for 1930.

The fuselage framework was built up of welded chrome-moly steel tubing in a rigid truss form, faired to shape with wooden formers and fairing strips then fabric covered. There was a large cockpit door on the right side with a convenient step provided for exit or entry; a small baggage compartment with maximum allowance for 25 lbs. was located forward of the instrument board and was accessible only from the outside through a hinged panel door. The side by side seating was rather chummy but still quite comfortable; the interior was upholstered and seat cushions were removable to allow for parachute packs. The semi-cantilever wing framework, in two halves, was built up of heavy-sectioned solid spruce spar beams that were "taper routed" for lightness; truss-type wing ribs were built up of spruce with plywood gussets, the leading edge was covered with dural sheet, the ailerons were a steel tube framework, and the completed wing panels were covered with fabric. A 12 gal. gravity-feed fuel tank was mounted in the root end of each wing half and a small sky-light was provided over the cockpit from the rear spar back. Robust N-type struts braced the wing to the fuselage over a pyramidal cabane of center-section struts and the out-rigger landing gear of 87 inch tread was built into a rigid truss with the wing-brace struts; shock absorbers were oil-spring struts, wheels were 24x4, and no wheel brakes were provided. The rubber-snubbed tail skid, mounted into lower end of rudder, was steerable in a limited range to help in ground maneuvering. Dual joy-stick controls were provided with the right hand stick quickly removable when not needed. The fabric covered tail-group was built up of welded chrome-moly steel tubing; the vertical fin was ground adjustable and the horizontal stabilizer was adjustable in flight. A wooden "Supreme" propeller was standard equipment. The next development in the Inland monoplane was the Warner powered "Super Sport" as discussed in the chapter for ATC # 315.

Listed below are Inland "Sport" model S-300 entries as gleaned from various registration records:

X-7255; Model S-300 (# S-301) Le Blond 60.
NC-9416;   „   (# S-302)   „
     ;   „   (# S-303)   „
NC-8087;   „   (# S-304)   „
NR-8088;   „   (# S-305)   „
NC-895H;   „   (# S-306)   „
NC-896H;   „   (# S-307)   „
NC-252K;   „   (# S-308)   „
NC-253K;   „   (# S-309)   „
NC-573K;   „,   (# S-310)   „
NC-574K;   „   (# S-311)   „
NC-252N;   „   (# S-312)   „
NC-253N;   „   (#-S-313)   „
NC-254N;   „   (# S-314)   „
NC-255N;   „   (# S-315)   „
NC-256N;   „   (# S-316)   „
NR-448V;   „   (# W-516-S)   „

Registration numbers for serial # S-300 and # S-303 unknown; serial # S-305 also as model S-300-U and later as model W-500; serial # S-312 later as # W-512; serial # S-308, S-314, S-315, S-316 later as model S-300-E with 70 h.p. LeBlond 5-DE; serial # W-516-S first as model W-500 later modified to model S-300.

Fig. 191. Trim S-300 was very popular for training or sport.

A.T.C. #260
(10-30-29)
KEYSTONE "PATRICIAN", K-78D

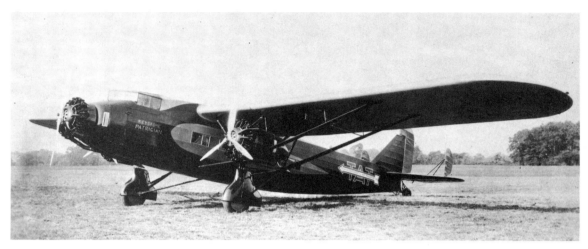

Fig. 192. Keystone "Patrician" model K-78D was the ultimate in passenger transport.

Everybody remembers the old "Ford" tri-motor and most everybody remembers the "Fokker" tri-motors but the number is few who remember the Keystone "Patrician". A huge tri-motored monoplane of truly majestic proportions, the stately "Patrician" was the last-word in air transportation and the largest and the fastest of the tri-engined air-liners in the world during this period. Assuming that passenger flights would continue to show steady increases, as they had just previous, and assuming that cargo tonnage would also continue its up-swing, Keystone designed the "Patrician" in an ideal size to accomodate the increased seating that would soon be required and also the added capacity that would be needed for air-borne cargo. In an accelerated program the prototype "Patrician" K-78 was formally introduced in late 1928 and soon embarked on a shake-down trip which was to take it across the country from New York to the Pacific coast and return; the promotional value of this trip was quite rewarding and the work-out for the new air-liner suggested several minor changes. To keep close watch on its further development, this first "Patrician" was placed in charter service with Colonial Air Transport on a run from New York to Boston. With government approval secured in the latter part of 1929, a production run of 10 airplanes had already been started for expected delivery to some of the large air-lines. Having more than a casual interest in the project, the Wright Aeronautical Corp.

purchased the first ship to leave the line as a flying test-bed for its new series "Cyclone" engines and equipped it as a plush "flying office"; the huge craft was fitted with private office rooms, a meeting room for company directors, and equipped with sleeping berths to accomodate passengers on lengthy trips. The second airplane was slated for service with Transcontinental Air Transport (T.A.T.) on their cross-country system but this "Patrician's" career was abruptly shortened by economic distress that developed in this country by 1930; orders for additional airplanes were soon cancelled. In an effort to stimulate some buying, Keystone lowered the price of the huge "Patrician" to a paltry $65,000. but it might as well have been a million dollars!

Certainly no stranger to aircraft manufacture, the Keystone Aircraft Corp., which stemmed from the early Huff-Daland Airplane Co., was organized in March of 1927 and kept itself busy building bomber airplanes and trainers for the U.S. government. Outside of the "Pathfinder" biplane transport, the "Patrician" was Keystone's first serious venture into commercial aircraft production. Influenced to a great extent by Curtiss and Wright interests, 3 "Cyclone" engines were chosen to power this huge air-liner and its construction was arranged in a pattern of materials quite familiar to Keystone Aircraft in its years of manufacturing aircraft for the military services. Though a distinct departure from the huge biplanes that were normally built, the graceful "Patrician" bore a marked resemblance to the "Panther" bombers in many ways. As is most always the case, the prototype "Patrician" was lighter, faster, and delivered a somewhat better performance than production versions that came out later; a beefed-up structure and improvements to the interior accounted for an increase of some 1300 lbs. to the empty weight and an increase in payload boosted the gross weight still further. The pilots cabin was extensively re-designed for better vision, engine nacelles were modified for better air-flow, and area was added to the tail-group for better control. The new series "Cyclone" engines were well muffled by large volume collector-rings and long tail-pipes to allow normal conversation in the cabin; with large windows running the full length of the interior, all passengers were treated to a panoramic view from their seats in the cabin. With calculations based on first cost, maintenance, and operation, the 18 passenger "Patrician" promised to be a profitable air-liner, providing each flight was seated to capacity.

The Keystone "Patrician" model K-78D was a high winged semi-cantilever monoplane with commodius seating for 18 passengers and a crew of two. Powered with three 9 cyl. Wright "Cyclone" engines of 525 h.p. each,

Fig. 193. "Patrician" was largest tri-motor transport at this time.

the total of 1575 h.p. allowed a good margin for power reserve and provided a performance that was the best of its kind for a ship of this type. Because of its scarcity not much lore has been recorded by this airplane but company releases indicated that it flew and handled well and was liked by the pilots. The prototype "Patrician" was certificated on a Group 2 approval numbered 2-85 issued 6-24-29 and production versions received a type certificate number on 10-30-29 for a 16,000 lb. gross and later amended to allow a 16,600 lb. gross weight. Records seem to disclose that only 3 examples of the "Patrician" were manufactured by the Keystone Aircraft Corp. at Bristol, Pa.; a division of the Curtiss-Wright Corp. Edgar N. Gott was the president; C. T. Porter was V. P. in charge of engineering; Albert P. Loening (brother of Grover Loening) was V.P. in charge of sales; C. L. Roloson was production manager; Stanley W. Jacques and K. G. Fraser were test pilots.

Fig. 194. Huff-Daland "Duster" of 1926 with Wright J4 engine; ancestor of Keystone line.

Fig. 195. Comfortable interior of "Patrician" had panoramic view.

Listed below are specifications and performance data for the Keystone "Patrician" model K-78D as powered with 3 Wright "Cyclone" (R-1750) engines of 525 h.p. each; length overall 61'7"; hite overall 13'0"; wing span 86'6"; wing chord 144"; total wing area 930 sq. ft.; airfoil Goettingen 398; wt. empty 10,224 lbs.; useful load 6376; payload with 360 gal. normal fuel was 3576 lbs.; payload with 420 gal. max. fuel was 3216 lbs.; gross wt. 16,600 lbs.; max. speed 144; cruising speed 120; landing speed 60; climb 1080 ft. first min. at sea level; ceiling 14,500 ft.; gas cap. normal 360 gal.; gas cap. max. 420 gal.; oil cap. 30-36 gal.; cruising range at 84 gal. per hour was 480-550 miles; price first quoted at $90,000., lowered to $85,000. and finally lowered to $65,000. in 1930. For comparison, the following figures are for the K-78 prototype; length 63'0"; hite 13'4"; wing span 88'0"; wing chord 12'; wing area 930 sq.ft.; wt. empty 8900 lb.; useful load 6100; payload with 420 gal. fuel 2980; gross wt. 15,000 lbs.; max. speed 151; cruising speed 130; landing speed 58; climb 1400 ft. first min. at sea level; ceiling 17,400 ft.; gas cap. 420 gal.; oil cap. 36 gal.; range 520 miles; price at the factory quoted as $90,000. The "Patrician" was also available with 3 Pratt & Whitney "Hornet" engines of 525 h.p. each.

The fuselage framework was an intricate structure that was built up of welded chrome-moly steel tubing, faired to a well-rounded shape with former bulkheads and fairing strips of aluminum alloy, then fabric covered. The main cabin of 20x6x6 foot dimension seated 18 with chairs arranged in two rows with an aisle between; there were 6 single seats

on the left side and six double seats on the right side. A lavatory was aft of the cabin area and a baggage compartment for 600 lbs. was beyond that. The thick cabin walls were sound-proofed and insulated and windows of shatter-proof glass were the full length of the cabin; there was a large door on each side aft for exit and entry. The cabin was tastefully appointed with numerous passenger conveniences and special interior arrangements were to be available on order. The semi-cantilever wing framework was built up in 4 sections; the inner wing panels from fuselage attach point to engine nacelles were built up of spar beams that were trusses of welded steel tubing, with truss-type metal wing ribs. The two outer wing panels were built up of spruce and plywood box-type spar beams and wooden truss-type wing ribs; the leading edges were covered with dural sheet and the completed framework of all 4 panels was covered in fabric. Fuel tanks were mounted in each inner panel of the wing and an oil tank was mounted in each engine nacelle. The huge wing was braced by two parallel struts that were fastened to a pyramidal structure containing the engine nacelles; the landing gear was braced in a connection with the wing-brace struts and the engine mountings to make a rigid truss structure. The landing gear of 19 ft. 6 in. tread used oleo & spring shock absorbing struts and was neatly faired by unusual wheel streamlines that were in the form of "spats"; wheels were 44x10 and Bendix brakes were standard equipment. The fabric covered tail-group was built up of welded chrome-moly steel tubing; two small sub-rudders provided additional fin area and the horizontal stabilizer was adjustable in flight. Metal propellers, navigation lights, tail wheel (oleo-spring), and hand-crank inertia-type engine starters were provided as standard equipment. The next Keystone development was the model K-85 amphibian as described in the chapter for ATC 395.

Listed below are Keystone "Patrician" entries as gleaned from registration records:

X-7962; Patrician (# 186) 3 Cyclone
X-98N;   „   (# 205)   „
NC-10N;   „   (# 206)   „

Serial # 186 was on Group 2 approval numbered 2-85.

A.T.C. #261
(10-24-29)
CURTISS-ROBERTSON "THRUSH", MODEL J

Fig. 196. Curtiss-Robertson "Thrush" model J with 225 H.P. Wright J6 engine.

The "Thrush" in the model J version was first developed into its present form by the Curtiss Aeroplane & Motor Co. and after tests had proven the modification satisfactory, the design was released to the Curtiss-Robertson division for manufacture. As discussed earlier, the "Challenger" powered version was sorely lacking in performance reserve and it took the installation of a Wright J6 "Seven" engine of 225 h.p. to bring forth the performance that would be expected from a craft of this size and type. The "Thrush" was now an airplane of good performance, and it had the ability to carry a sizeable payload with apparent ease.; its short-field performance was one of its better features and it was not uncommon to see them operating from pasture-airports throughout the country. Curtiss had a demonstration team operating in several countries of the Orient to stimulate export sales; 15 of the "Thrush" model J were reported ordered by the Chinese government for a proposed route, with air-mail and passenger service between Hankow, Shanghai, Nanking, Peking, and Canton. Information whether this service was instituted or whether the total of 15 aircraft were ever delivered, could not be found. Other examples of the Curtiss-Robertson "Thrush" were operating in this country doing various chores, and the model J soon became known as a willing and able jack of all trades.

The Curtiss-Robertson "Thrush" model J, as powered with the Wright J6 (R-760) engine of 225 h.p., was a buxom and boxy high wing cabin monoplane of rather large proportions; it had more than ample room for six, with interior appointments that were sensible and practical. Dual controls were provided and all movable surfaces were aerodynamically balanced to lighten the forces at the "stick"; most reports would indicate that the big "Thrush" was easy to fly and quite maneuverable for a ship of this size. Unhampered entry was gained with the help of conveniently placed steps and through two large doors on the right hand side; muffling of the engine kept the noise level low and an extremely long tail-pipe directed gases away from the cabin area. As can be guessed from descriptions given, the "Thrush" was primarily designed as a light transport to serve on feeder-lines, or operate as air-taxi in chartered service; that it was not built in greater number can only be blamed on the fact that it appeared at a time during late 1929 and early 1930 when the aircraft industry was rapidly losing its market. The Curtiss-Robertson "Thrush" model J, as powered with the Wright J6 "Whirlwind Seven" of 225 h.p., received its type certificate number on 10-24-29 (approval was actually based on the specifications of ATC # 236), and some 10 examples of this model were manufactured

Fig. 197. Load carrying ability and short field performance of "Thrush" adaptable to many uses.

by the Curtiss-Robertson Airplane Mfg. Co. at St. Louis (Anglum), Mo. Maj. Wm. B. Robertson was the president; Ralph Damon was in charge of production; and likeable Dale "Red" Jackson was the chief pilot in charge of test and development.

Listed below are specifications and performance data for the Curtiss-Robertson "Thrush" model J as powered with the Wright J6 engine of 225 h.p.; length overall 32'7"; hite overall 9'3"; wing span 48'0"; wing chord 84"; total wing area 305 sq.ft.; airfoil "Curtiss C-72"; wt. empty 2260; useful load 1540; payload with 60 gal. fuel was 970 lbs.; payload with 110 gal. fuel was 670 lbs.; gross wt. 3800 lbs.; max. speed 122; cruising speed 104; landing speed 52; climb 650 ft. first min. at sea level; climb to 5000 ft. in 9.4 min.; climb in 10 min. was 5300 ft.; service ceiling 13,200 ft.; gas cap. normal 60 gal.; gas cap. max. 110 gal.; oil cap. 5-9 gal.; cruising range normal fuel was 490 miles; range with max. fuel was 900 miles; price at the factory field was $12,000.

Following traditional Curtiss practice, the fuselage framework was built up of seamless duralumin tubing and fastened together with duralumin wrap-around fittings and dural hollow rivets; some of the more highly stressed portions of the framework were built up of welded chrome-moly steel tubing to beef-up the structure. The completed fuselage framework was lightly faired to shape and fabric covered; useable cabin area measured 38x81x41 inches and seats were easily removable for the hauling of bulky cargo. The semi-cantilever wing framework was built up of solid spruce spar beams and stamped-out aluminum alloy wing ribs; the leading edge was covered with aluminum alloy sheet, fuel tanks were mounted in the wing and the completed framework was covered in fabric. Wing braces were parallel struts of chrome-moly steel tubing in streamlined section, a similar arrangement to that used on the smaller "Robin"; the out-rigger type landing gear of 116 in. tread was tied into the wing struts as a robust truss. Wheels were 30x5 or 32x6 and Bendix brakes were standard equipment; a steerable tail wheel was fitted to make for better ground maneuvering. The fabric covered tail group was built up of welded chrome-moly steel tubing and all movable surfaces were aerodynamically balanced by projecting "horns"; the fin was ground adjustable to offset the engine torque and the horizontal stabilizer was adjustable in flight. Dual controls, a Curtiss-Reed metal propeller, an inertia-type engine starter, and wiring for navigation lights, were standard equipment. The next Curtiss-Robertson development was the rare "Robin W" that was powered with the Warner "Scarab" engine; for a discussion of this craft refer to the chapter for ATC # 268 in this volume.

Listed below are "Thrush" model J entries that were gleaned from registration records; this appears to be the total production:

NC-522N;	Thrush J	(# 1001)	Wright J6-7-225
NC-523N;	„	(# 1002)	„
NC-542N;	„	(# 1003)	„
NC-552N;	„	(# 1004)	„
NC-553N;	„	(# 1005)	„
;	„	(# 1006)	„
NC-562N;	„	(# 1007)	„
NC-580N;	„	(# 1008)	„
NC-581N;	„	(# 1009)	„
NC-582N;	„	(# 1010)	„

Serial # 1005 also as "J Special" on Group 2 approval numbered 2-210; registration number for serial # 1006 unknown, may have been exported; serial # 1010 also eligible as "J special" with Wright J5 engine; the "Thrush" was also tested with Wright J6 of 300 h.p., but none were scheduled for production.

Fig. 198. Plush B-7 with 420 H.P. Wasp engine was last of the Ryan "Brougham."

Starting from the days of the ocean-crossing "Spirit of St. Louis", back in 1927, the Ryan "Brougham" series had progressed up through the model B-1, the B-3, and the B-5; each model change gained a slight improvement in utility and performance for the lady-like "Brougham", until it reached the point where it was considered one of the best craft of its type. Of medium capacity and very versatile as to environment or chore, the "Brougham" served on many of the smaller air-lines, performed a multitude of duties with the flying services around the country and was one of the first to prove to business the usefulness and advantages to be gained by air-travel. The many advancements incorporated into the model B-5 seemed like the end of the line for this particular series, outside of perhaps a few minor details and it was all that one could ask from a ship of this type, but customer demands have a way of influencing the final trend of things, so the model B-7 was designed as the end result.

The Ryan "Brougham" model B-7 was actually a more buxom and slightly enlarged version of the popular B-5; to provide the extra performance that would be expected from this new model, it was powered with the 9 cyl. Pratt & Whitney "Wasp" C-1 engine of 420 h.p. Though tailored to the demands of a particular clientele, the model B-7 was still pretty much of a "Brougham" in appearance and its general arrangement; the slightly longer fuselage was faired out a little deeper to blend in with the bigger engine and the tail-group was enlarged for better directional control. All-round performance was increased considerably and the impeccable appointments provided an elegance to suit the most demanding customer. Harry C. Williams, an air-minded millionaire sportsman from New Orleans, who was also the vice-president of the Wedell-Williams Air Service, used a B-7 as a personal transport for pleasure jaunts to hunting and fishing grounds, or for business promotion. Other examples of the B-7 were serving in various lines of business or on fast schedules with feeder-lines.

The model M-1 which was jointly designed by Wm. Waterhouse, W. A. "Bill" Mankey, and Hawley Bowlus, was Ryan's first offering to commercial aviation; there was certainly quite a difference in only 4 short years between the stark utility of the M-1 of 1926 and the graceful elegance of the B-7 in 1930. The type certificate number for the 6 place Ryan "Brougham" model B-7, as powered with the 420 h.p. "Wasp" engine, was issued 10-26-29 for both landplane and seaplane versions; some 8 examples of this model were manufactured by the Ryan Aircraft Corp. at the Lambert-St. Louis Airport in Robertson, Mo.; a division of the Detroit Aircraft Corp. An interesting side-light at the Ryan plant was the fact that women had entirely replaced the men by now in the wing and fuselage covering department; it had been found that the girls handled the fabric better and performed a far neater job. Sometime in 1930 manufacture of the B-7 was transferred to the plant in Detroit, Mich.

Listed below are specifications and performance data for the Ryan "Brougham" model B-7 as powered with the 420 h.p. "Wasp" engine; length overall 29'11"; hite overall 8'10"; wing span 42'4"; wing chord 84"; total wing area 280 sq.ft.; airfoil Clark Y; wt. empty 2503 lbs.; useful load 1780; payload with 100 gal. fuel was 950 lbs.; gross wt. 4283 lbs.; max. speed 150; cruising speed 128; landing speed 60; climb 1350 ft. first min. at sea level; ceiling 20,000 ft.; gas cap. 100 gal.; oil cap. 8 gal.; cruising range at 23 gal.

per hour was 500 miles; price at the factory field was $16,985. The following figures are for seaplane version as fitted with Edo K twin-float gear; wt. empty 3053; useful load 1617; payload with 100 gal. fuel 787 lbs.; with max. fuel payload was limited to 4 pass. & 120 lbs. baggage; gross wt. 4670; max. speed 138; cruising speed 120; landing speed 65; climb 1150 ft. first min. at sea level; ceiling 19,000 ft.; price at factory was approx. $19,000. Gross weight of landplane was later amended to allow 4300 lbs., also eligible with 120 gal. fuel.

The fuselage framework was built up of welded chrome-moly steel tubing in a rigid truss form, deeply faired to shape with wooden formers and fairing strips then fabric covered. The walls of the cabin were heavily sound-proofed and insulated and the interior which measured 114x47x47 inches was richly upholstered in blue mohair fabric with polished mahogany trim; a baggage compartment of 20 cu. ft. capacity with allowance for 100-125 lbs. was aft of the cabin section and was accessible from either inside or out. The cabin had provisions for ventilation and exhaust-manifold heat; large windows offered excellent visibility to passengers and lower side-panels offered ample vision for the pilot. A convenient step and a large door on each side offered unhampered entry or exit; passenger seats were of the bench-type or individual bucket-type on order. The 2 front seats were of the bucket-type and the right

Fig. 199. Ryan B-7 served in business and small air-lines.

front seat could be folded up and pilot's seat slid over to a central position, if so desired; dual controls were of the joy-stick type. The semi-cantilever wing framework was built up of 3-ply laminated spruce spars that were routed out to an I-beam section, with wing ribs built up of spruce with mahogany ply-wood gussets in a Warren type truss; the leading edge was covered with dural sheet and the completed framework was covered in fabric. The wing bracing struts were of heavy gauge chrome-moly steel tubing and were encased in balsa-wood fairings that were shaped to an airfoil form for added lift and stability; the two fuel tanks were mounted in the wing flanking the fuselage. The split-axle landing gear of 120 inch tread was of the out-rigger type and was formed into a strong truss with the wing bracing struts; shock absorbers were "Aerol" (air-oil) struts, wheels were 32x6 and Bendix brakes were standard equipment. The fabric covered tail-group was built up of welded chrome-moly steel tubing; the fin was ground adjustable and the hori-zontal stabilizer was adjustable in flight. The engine's roar was well muffled by a large volume collector-ring and a long tail-pipe;

an oil-cooler was provided to regulate oil temperature. A metal propeller, navigation lights, full-swivel tail wheel, and Eclipse in-ertia-type engine starter were standard equip-ment. A low-drag "Townend ring" engine cowling was offered as optional equipment as were larger fuel tanks to allow 120 gal. fuel capacity. The model B-7 was the last and the finest example in the "Brougham" series; the next Ryan development was the smaller "Foursome" as described in the chapter for ATC # 346.

Listed below are "Brougham" model B-7 entries as gleaned from registration records:

NC-549N; Ryan B-7 (# 249) Wasp 420.
NC-555N;   "   (# 250)   "
NC-724M;   "   (# 251)   "
NC-720M;   "   (# 252)   "
NC-721M;   "   (# 253)   "
     ;   "   (# 254)   "
NC-723M;   "   (# 255)   "
     ;   "   (# 256)   "

Registration numbers for serial # 254 and # 256 unknown; serial # 255 later on Group 2 approval 2-223 with 132 gal. fuel.

Fig. 200. Mohawk "Pinto" model M-1-CK with Kinner K5 engine; ideal for sport.

When the original Mohawk "Pinto", as designed by Wallace C. "Chet" Cummings, was first introduced, the stubby little craft created quite a stir among the flying folks and was discussed with very much interest and excitement; not before too long; the sassy "Pinto" had earned a reputation certainly not conducive to overwhelming popularity. Low-winged light monoplanes were still looked at with a very skeptical eye and the capricious traits attributed to the early Mohawk monoplane were enough to cause uneasiness to all but the stout of heart. It is true that the early Warner-powered "Pinto" was jokingly likened to "having a tiger by the tail" but a fair amount of understanding and a determined hand on the stick furnished a performance that was a joyful experience. Called in to teach this "playful brat" some manners, Prof. John D. Akerman of the Univ. of Minnesota tamed the "Pinto" a good deal by redesigning its aerodynamic arrangement and this much be-rated craft finally lived down its earlier reputation and proved it to be largely of talk-promoted fallacy. Now the "Pinto" was judged entirely on its inherent merits and not pre-judged by any careless hangar-talk.

First brought out as the "Spurwing" and exhibited at the Detroit Air Show of 1929, the new series "Pinto" was more or less similar in all respects except for its aerodynamic arrangement; moment arms were generously lengthened to decrease its sensitivity and larger control surfaces provided a more positive control of the airplane in abnormal attitudes. A direct descendant of the "Spurwing", the model M-1-CK was also a low-winged cantilever monoplane seating two in tandem in open cockpits; power for this 1930 version was the 5 cyl. Kinner K5 engine of 100 h.p. Though relieved of some of its capricious habits, the new series "Pinto" was still quite lively in general behavior and possessed a staunch character that stood up quite well under abuse and hard usage. Basically designed as a sport-plane for the private-owner flyer, to allow interesting jaunts to out-of-way places or to satisfy the playful mood around the home port, the M-1-CK was also well suited for primary and secondary phases of pilot-training. One example of the M-1-CK was tested by the Air Corps for possible use as a primary trainer; it was designated the XPT-7. Tests were apparently satisfactory but no orders were placed for similar airplanes. The type certificate number for the Mohawk "Pinto" model M-1-CK as powered with the 100 h.p. Kinner engine, was issued 10-28-29 and some 4 or more examples of this model were manufactured by the Mohawk Aircraft Corp. at Minneapolis, Minn. Stanley Partridge was the president; A. S. Koch was V.P. and general manager; John D. Akerman was chief

Fig. 201. Army "Pinto" was XPT-7.

of engineering; and Hiram W. Sheridan was the chief pilot for a time.

Listed below are specifications and performance data for the Mohawk "Pinto" model M-1-CK as powered with the 100 h.p. Kinner K5 engine; length overall 24'2"; hite overall 7'7"; wing span 34'11"; wing chord at root 75"; wing chord at tip 34"; wing chord mean 53"; total wing area 145 sq.ft.; airfoil "Mohawk" # 3; wt. empty 1142; useful load 658; payload with 32 gal. fuel was 269 lbs.; gross wt. 1800 lbs.; (XPT-7 crew wt. 434 lbs.); max. speed 115; cruising speed 98; landing speed 40; climb 850 ft. first min. at sea level; ceiling 15,000 ft.; gas cap. 32 gal.; oil cap. 4 gal.; cruising range at 6.5 gal. per hour was 475 miles; price at the factory was announced as $6000. but soon lowered to $5300. and reduced to $4800. in May of 1930.

The fuselage framework was built up of welded chrome-moly steel tubing, faired to shape with wooden formers and fairing strips and fabric covered. The cockpits were 30 in. deep x 42 in. long and well protected from the weather by large windshields; dual joystick controls were provided with the front stick quickly removable when not in use. The fuel tank was mounted high in the

Fig. 202. Mohawk "Spurwing" was basis for new "Pinto" series.

fuselage just ahead of the front cockpit, and a fair-sized luggage compartment of 9.3 cu. ft. capacity was in the turtle-back section behind the rear cockpit. All controls were actuated by metal push-pull tubes, except for the rudder which was operated by stranded steel cable. The cantilever wing framework, in two halves, was built up of spruce and mahogany plywood box-type spar beams with wing ribs of spruce diagonals and flanges around mahogany plywood webs; the leading edges were covered with plywood sheet, the Friese-type ailerons were of a steel tube structure, and the completed framework was covered with fabric. The split-axle landing gear of 102 inch tread used "Aerol" (air-oil) shock absorbing struts; wheels were 26x4 and Bendix brakes were standard equipment. The landing gear attachment and its arrangement was similar to the early "Pinto", except that the oleo-legs were now on the inside of the wheels to promote quicker servicing. The fabric

covered tail group was built up of welded chrome-moly steel tubing; the fin was ground adjustable and the horizontal stabilizer was adjustable in flight. A Hartzell wood propeller, wiring for navigation lights, and a spring-leaf tail skid were standard equipment; a metal propeller, and Heywood air-operated engine starter were optional. The next development in the new "Pinto" series was the Warner-powered M-1-CW as described in the chapter for ATC # 297 in this volume.

Listed below are Mohawk "Pinto" model M-1-CK entries as gleaned from registration records:

NC-67N; Pinto M-1-CK (# 3) Kinner K5.
NC-180N;   „   (# 4)   „
NC-297V;   „   (# 5)   „
NC-319V;   „   (# 6)   „

Serial # 4 also as model M-1-CW; serial number for XPT-7 unknown.

Fig. 203. "Pinto" stance typical of its get-up-and-go.

Fig. 204. St. Louis "Cardinal" model C2-90 with 90 H.P. Le Blond engine.

First developed in 1928 and finally offered to the public in 1929 as the C2-60, the "Cardinal" monoplane was also offered in a higher-powered version called the model C2-90. Particularly leveled at the sportier type of pilot, the "Cardinal" C2-90 was powered with the new 7 cyl. LeBlond 7-D engine of 90 h.p. and performance was substantially improved over the lower-powered version which mounted an engine of 65 h.p. Typical to several other craft in the light cabin class, the "Cardinal" was also planned to offer several extra features not normally found in a ship of this type and was easily available in several power ranges to offer a performance to suit the whim or the need. Despite the fancied expectations, sales were not particularly plentiful in this type of aircraft, especially due to the economic sag of late 1929 and competition already entrenched was rather hard to buck; because of this, the C2-90 was selling rather slowly and was built only in a small number. The first "Cardinal" to come off the production line, a model C2-60, was ferried to the west coast to a dealer in Calif. and the second ship was added to the flight-line at the Von Hoffman Flying School in St. Louis; others were soon scattered thinly around the country. The 7 cyl. LeBlond engine had been just recently approved and the "Cardinal" monoplane was one of the first to utilize this powerplant; labeled the model 7-D, this engine developed

90 h.p. at 1975 r.p.m. and was a very compact power unit. Had aircraft manufacture continued on at a more normal level, it is very likely that this engine would have been seen on many more airplane models.

The St. Louis "Cardinal" model C2-90 was a light high-winged cabin monoplane that seated two side by side with good visibility and ample comfort; powered with the 90 h.p. Le-Blond engine, performance was a good bit better than was usually offered in an airplane of this type. Unlike the more popular airplanes that were built in much greater number, the "Cardinal" C2-90 left very little lore behind to describe its nature, so we can only assume that flight characteristics and its general behavior were about average for a ship of this type. It is quite likely that had the "Cardinal" any vicious or unfriendly habits, the word would have surely gotten around at hangar-sessions all over the country. Tested thoroughly in two prototype aircraft the "Cardinal" in the production version was a craft that offered several improvements; the tail-skid was moved further aft to the very end of the fuselage to act as a longer "wheelbase" and swiveling action of the skid permitted better ground maneuvering. Coupled with internal-expanding brakes of the company's own design, this little craft should have been quite manageable. The rudder was notched to allow clearance at the tail-skid and the "balance horn" on the rudder was apparently discarded to reduce the sensitivity. Due to the extra built-in features and the robust structure, the model C2-90 weighed in a bit more than the average for this type but this didn't seem to penalize the performance. The type certificate number for the "Cardinal" model C2-90, as powered with the LeBlond 7-D engine, was issued 11-14-29 and some 6 or more examples of this model were manufactured by the St. Louis Aircraft Corp., a subsidiary of the St. Louis Car Co. of St. Louis, Mo.

Listed below are specifications and performance data for the "Cardinal" model C2-90 as powered with the 90 h.p. LeBlond engine; length overall 20'6"; hite overall 7'0"; wing span 32'4"; wing chord 60"; total wing area 162 sq.ft.; airfoil Clark Y; wt. empty 999 lbs.; useful load 558; payload with 30 gal. fuel was 190 lbs.; gross wt. 1557 lbs.; max. speed 118; cruising speed 100; landing speed 40; climb 880 ft. first min. at sea level; ceiling 10,500 ft.; gas cap. 30 gal.; oil cap. 3 gal.; cruising range at 6 gal. per hour was 450 miles; price at the factory was first quoted as $3450., later raised to $3750. and then lowered to $3250. in mid-1930.

The construction details and general arrangement of the model C2-90 were typical to that of the model C2-60 as described in the chapter for ATC # 273 in this volume. Cabin dimensions of the "Cardinal" were 35 in. wide x 44 in. long and clear head-room was 39 in.; a large door and a convenient step were provided on the right side for exit and entry. Dual controls were joy-stick type and the brakes were operated by separate pedals; a small baggage compartment of 4 cu. ft. capacity with allowance for 15 lbs. at full gross load, was located in back of the seat. The split-axle landing gear of 75 inch tread used oil-draulic shock absorbing struts; wheels were wire-type with 26x4 clincher-type tires and wheel brakes were standard equipment. A Hartzell or Supreme wooden propeller and wiring for navigation lights were standard equipment; a metal propeller, navigation lights, and engine starter were optional. The first development in the "Cardinal" monoplane series was the model C2-60 as described in the chapter for ATC # 273 in this volume.

Listed below are "Cardinal" model C2-90 entries as gleaned from registration records: NC-989K; Model C2-90 (# 107) LeBlond 90. NC-761M;   „   (# 111)   „ NC-559N;   „   (# 113)   „ NC-560N;   „   (# 114)   „ NC-991K;   „   (# 115)   „ NC-760M;   „   (# 116)   „

Serial # 107 first as C2-60; serial # 113 later modified to model C2-85 with LeBlond 5-DF; serial # 103 originally as C2-60, rebuilt in recent years as a model C2-90.

Fig. 205. "Cardinal" prototype with Le Blond 60 Engine.

Fig. 206. "Cardinal" rebuilt to fly again in recent years.

A.T.C. #265
(11-2-29)
PARAMOUNT "CABINAIRE", 165

Fig. 207. Paramount "Cabinaire" model 165 with 5 cyl. Wright J6 engine.

The Paramount "Cabinaire" was an interesting example in a small family-type airplane; trim and very compact, it didn't have that big expensive look and seems to be a craft one would thoroughly enjoy owning and flying around the home port. Rather unusual too is the fact that the "Cabinaire" was an enclosed biplane and introduced at a time when the big switch was being made to monoplanes for a craft of this type. It doesn't take much imagination to see the underlying purpose for this particular selection of configuration and realize that it was perhaps guided by a desire for an abundance of wing area around a fuselage that was hardly much bigger than the average 3 place biplane of the open cockpit type. Consequently, the result of this selection is a petite 4 place airplane that is about 500 lbs. lighter than the average monoplane of this capacity with a performance on 165 h.p. that would otherwise not have been possible. Quite proud of the "Cabinaire" and rightfully so, Walter Carr demonstrated its ability on every opportunity; as a contestant in the hard-fought National Air Tour for 1930, he flew the "Cabinaire" model 165 to 15th place amongst a very determined field.

Always a firm believer in the merits of the biplane, Walter Carr was perhaps as avid an advocate for the two-winger as Clyde Cessna

and others were for the monoplane. It is true that the monoplane had characteristics that would win in the end but the biplane still had many advantages in its favor during this particular period of time. Considered as one of the smaller operations in the aircraft industry, Paramount Aircraft had not the large amounts of capital that some enjoyed so they had to operate with modest expenditures and with a modest working crew; had Paramount been blessed with more sales of their "Cabinaire" there is no telling what might have been. In development for over a year, the "Cabinaire" had the misfortune of being introduced on the threshold of the "depression" and this was a bitter-pill that many found hard to swallow. Reaching its peak of popularity along about this time, the cabin biplane soon began fading from the scene and except for the popular "Waco" cabin biplanes and the fabulous Beech "Stagger-Wing" was almost unheard of again.

Walter J. Carr, the designer of the "Cabinaire" biplane series, was a pioneer pilot that was well known in Michigan flying circles with exhibition flights that date back to at least 1915. Typical of many old-timers who saw plenty of room for improvement in the old crates they were then flying, Carr took a fling at building airplanes and designed the "Maiden Saginaw" in 1924. Also interested in promoting the use of air transportation, Carr became chief of operations and chief pilot for Northern Airways which maintained a route from Detroit to Saginaw and Bay City. Versatile and qualified through experience in many phases of aviation, Walter Carr undertook to test-fly the first Warner "Scarab" engine in 1927 which had been mounted in his own "Travel Air" biplane. With 156 hours of flying time recorded on this first Warner engine (Serial # 1), tests were pronounced satisfactory, the engine received its government approval and Carr fell heir to this first engine which he also later used in his first "Cabinaire" type. Stories vary to some extent as stories will, but from close examination of the prototype airplane in this series it is easy to believe that Walter Carr's own "Travel Air" was the back-bone for his first "Cabinaire", right down to the "elephant ear" ailerons and all. We could say that the "Travel Air" was added to here and there and modified into a "Cabinaire". With flight tests performed and many lessons learned, the next "Cabinaire" though still very similar was redesigned in a personality all its own and in a craft that harbored many innovations. Organized late in 1928, Paramount Aircraft had their "Cabinaire" series in development for nearly a year before receiving government approval for manufacture.

The Paramount "Cabinaire" model 165 was a cabin biplane of rather petite proportion that seated 4 in fair comfort but in chummy proximity. Powered with the 5 cyl. Wright J6 engine of 165 h.p. performance compared favorably to the very best and economy per seat-mile was one of its cardinal features. Planned especially for the private owner or the low-budget business man, the "Cabinaire" cabin biplane was designed as an ideal family-type airplane or an economical air-taxi that could operate cheaply and easily in and out of smaller airports adjacent to the smaller towns, towns that one would find in out of way places and not be limited to the well-developed airfields in the bigger cities. The first batch of "Cabinaires" were powered with the 110 h.p. Warner engine and it is surprising the performance that was available from this 4-seated craft; well arranged aerodynamically its behavior was positive and enjoyable. To compensate for some losses in speed due to bracing struts and rigging, Carr made use of the N.A.C.A. type low-drag engine cowling to cancel out the losses; the "Cabinaire" was one of the first few to take advantage of this type of engine fairing. The "Cabinaire 165" with its bigger Wright engine was not particularly adaptable to the N.A.C.A. cowling because of its larger diameter but the added horsepower compensated for drag losses that were more critical with lower power. The type certificate number for the "Cabinaire" model 165 as powered with the 165 h.p. Wright J6 (R-540) engine was issued 11-2-29 and only one example of this model was manufactured by the Paramount Aircraft

Fig. 208. Cabinaire offered practical utility and high performance on nominal horsepower.

Corp. at Saginaw, Mich. J. E. Behse (or Behre) was president, treasurer, and sales manager; and Walter J. Carr was V.P., plant manager, chief design-engineer, and chief pilot in charge of test and development.

Listed below are specifications and performance data for the Paramount "Cabinaire 165" as powered with the 165 h.p. Wright J6 engine; length overall 24'7"; hite overall 9'0"; wing span upper 33'0"; wing span lower 29'0"; wing chord upper 66"; wing chord lower 56"; wing area upper 173 sq.ft.; wing area lower 136 sq.ft.; total wing area 309 sq.ft.; airfoil "Carr"; wt. empty 1620 lbs.; useful load 1010; payload with 50 gal. fuel was 510 lbs.; gross wt. 2630 lbs.; max. speed 120; cruising speed 102; landing speed 45; climb 780 ft. first min. at sea level; ceiling 12,000 ft.; gas cap. 50 gal.; oil cap. 4 gal.; cruising range at 9.5 gal. per hour was 500 miles; price at the factory was $7500., lowered to $6750., and cut to $5750. in June of 1931.

The fuselage framework was built up of welded chrome-moly steel tubing, faired to shape with metal formers and wooden fairing strips then fabric covered. The chummy cabin of 83 cu. ft. volume capacity seated 4 with two individual seats in front and a bench type seat in back; a door on right side front provided entry to the front seats and a door on left side rear provided entry to the seat in back. Entry to the cabin was an easy step from the wing-walk. Upholstery was plain and practical with large windows for ample visibility. The wing framework was built up of laminated spruce spar beams with spruce and plywood truss-type wing ribs; the completed framework was covered with fabric. Fuel tanks were mounted in the root end of each upper wing panel, outboard from the center-section which was supported atop the fuselage by two small N-type struts. The wide tread landing gear was of the rugged out-rigger type with "Aerol" (air-oil) shock absorbing struts; wheels were 30x5 and Bendix brakes were standard equipment. The "Cabinaire" was planned to operate on "Edo" floats but was not approved. The wheel brakes were operated by depressing a patented attachment

Fig. 209. Early "Cabinaire" carried four on 110 H.P.; neat installation of N.A.C.A. cowling contributed heavily to performance.

on the control wheel with slight pressure of the thumbs; the spring-leaf tail skid had a patented revolving "dural ball" for use on hard-surface runways. The fabric covered tail-group was built up of welded steel tubing; the fin was ground adjustable and the horizontal stabilizer was adjustable in flight. Balanced-hinge ailerons (Friese type) and a balanced rudder were provided to lighten control loads. A metal propeller, wheel brakes, and wiring for navigation lights were standard equipment. Navigation lights and inertia-type engine starter were optional. The next development in the "Cabinaire" series was the model A-70 as powered with the 165 h.p. Continental engine; a Group 2 approval numbered 2-233 was issued 7-9-30. Struggling through part of the economic depression, Paramount found it difficult to continue and finally closed its doors; in June of 1932 manufacturing rights for the "Cabinaire" were offered for sale.

Listed below is the only known entry for the "Cabinaire 165":
NC-17M; Cabinaire 165 (# 7) Wright R-540.

Serial # 1 thru # 5 were on Group 2 approval 2-165 as 3 place with Warner 110; serial # 6 was on Group 2 approval 2-164 as 4 place with Warner 110; identity of serial # 8 unknown; serial # 9 was "Cabinaire A-70" with 165 h.p. Continental engine on Group 2 approval 2-233; X-4254 was prototype for "Cabinaire" series.

A.T.C. #266
(11-14-29)
CURTISS "FLEDGLING", MODEL J-1

Fig. 210. Curtiss Fledgling model J-1 with 165 H.P. Wright J6 engine.

The "Curtiss Fledgling", though a modernized trainer for the times in every respect, actually had more wing struts and nearly as many bracing wires as the old Curtiss "Jenny" (JN-4D) of some 10 years earlier! It is certainly well apparent at a glance that the designers of the "Fledgling" trainer were not greatly concerned with parasitic resistance, but strived instead to create an airplane that would be rugged of structure, easy to service when need be, and serve best in the capacity of teaching student-pilots how to fly an airplane. From records and lore left behind by this airplane, we cannot help but agree that the Curtiss "Fledgling" was one of the finest training airplanes ever built, and one that held a special reverance in the hearts of thousands of pilots who first wobbled their way through the sky with the helping hand of this good-natured airplane.

Though the greatest number of Curtiss "Fledglings" by far, were powered with the 6 cyl. Curtiss "Challenger" engine of 170 h.p., several other versions were built at different times. One of these was the model J-1 which

was a typical 2 place open cockpit "Fledgling" trainer in all respects, except for the engine installation which in this case was the 5 cyl. Wright J6 of 165 h.p. The Wright powered model J-1 had slightly less horsepower but the difference in performance was hardly noticeable; it still retained all of the admirable features and characteristics of nature that made the "Fledgling" such a great favorite at the 40 different Curtiss Flying Service air-stations located across the country. The Curtiss "Fledgling" model J-1 as powered with the 5 cyl. Wright J6 (R-540) engine, received its type certificate number on 11-14-29 and some 4 or more examples of this model were built by the Curtiss Aeroplane & Motor Co. in one of its plants at Buffalo, New York. All design work and test development for the "Fledgling" series was performed by the Curtiss corps of engineers, headed by T. P. Wright, at the parent-plant in Garden City, L.I., N.Y.

Listed below are specifications and performance data for the Curtiss "Fledgling" model J-1 as powered with the 5 cyl. Wright J6 engine of 165 h.p.; length overall 27'10"; hite overall 10'4"; wing span upper 39'1"; wing span lower 39'5"; wing chord both 60"; wing area upper 188 sq.ft.; wing area lower 177 sq.ft.; total wing area 365 sq.ft.; airfoil Curtiss C-72; wt. empty 2009; useful load 691; crew wt. 380 lbs.; gross wt. 2700 lbs.; max. speed 102; cruising speed 87; landing speed 45; climb 620 ft. first min. at sea level; climb in 10 min. was 4430 ft.; ceiling 11,650 ft.; gas cap. 40 gal.; oil cap. 3.5 gal.; cruising range at 9.2 gal. per hour was 390 miles; crew wt. includes 2 parachutes at 20 lbs. each; baggage allowance was 43 lbs. The "Fledgling Jr." with single bay wing cellule was also tested with the 5 cyl. Wright J6 engine of 165 h.p.; the following is submitted to offer comparison. Length overall 28'3"; hite overall 10'4; wing span both 31'6"; wing chord both 60"; total wing area 289 sq.ft.; wt. empty 1874; useful load 646; crew wt. 380; gross wt. 2520 lbs.; max. speed 106; cruising speed 88; climb 575 ft.; ceiling 10,750 ft.; climb in 10 min. was 4650 ft.; gas cap. 40 gal.; oil cap. 3.5 gal.; range 405 miles; from a study of the specs presented here one could surmise that no distinct advantage was gained by resorting to short-span wings of single bay construction on the "Fledgling" design.

The fuselage framework was built up of welded chrome-moly steel tubing in a Warren truss form, faired to shape and fabric covered. The seats were adjustable and had deep wells to fit a parachute pack; a small tool locker was located in the turtleback behind the rear seat. The wing framework was built up of solid spruce spar beams that were routed out for lightness, with spruce and plywood wing ribs; the leading edge of the upper wing was covered in aluminum alloy sheet and the leading edge of the lower wing was covered in plywood sheet to ward off dents caused by flying debris on take-off. The aileron framework was built up of welded steel tubing and the completed wing framework was covered in fabric; the 4 ailerons, connected together in pairs by a streamlined push-pull strut, were operated independently by metal push-pull tubes. The balanced rudder and elevators were operated by stranded steel cable. The fuel tank of 40 gal. capacity was mounted high in the fuselage just ahead of the front cockpit; an oil tank of 3.5 gal. capacity was mounted on the firewall of the engine compartment. The split-axle landing gear of 87 inch tread was of two long telescopic legs, using a combination of oleo struts with rubber rings in compression as shock absorbers; the wheels were 28x4 and no brakes were provided. The fabric covered tail-group was built up of welded chrome-moly steel tubing; the fin was ground adjustable and the horizontal stabilizer was adjustable in flight. To promote easier ground maneuvering without brakes, the tail-skid was steerable. All points requiring frequent inspection or servicing were easily accessible through

Fig. 211. Fledgling J-1 had just as many struts and wires but was rare version in the series.

numerous removable panels; all points of wear were provided with greasing or oiling fittings. A Curtiss-Reed metal propeller, and hand-crank inertia-type engine starter were standard equipment. For a more complete account of the "Fledgling" structure, maintenance, and flight characteristics, refer to details described in the chapters for ATC # 182 and # 191 of U.S. CIVIL AIRCRAFT, Vol. 2. For the next development in the "Fledgling" trainer series, refer to the chapter for ATC # 269 in this volume which discusses the 225 h.p. Wright powered model J-2.

Listed below are "Fledgling" model J-1 entries that were gleaned from registration records:

NC-8663; Fledgling J-1 (# B-5)
　　　　　　　　　　　　　　Wright J6-5-165
NC-8677;　　　　,,　　　(# B-18)　　,,
NX-8690;　　　　,,　　　(# B-21)　　,,
NC-8691;　　　　,,　　　(# B-22)　　,,

Serial # B-22 later on Group 2 approval numbered 2-472 as J-1 Special with Wright J5 engine.

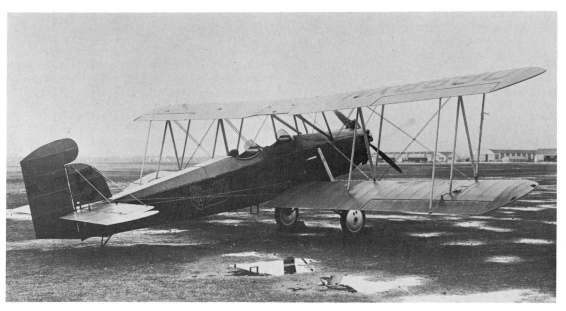

Fig. 212. What the Fledgling had in bulk and wing area it matched in patience and good nature.

VERVILLE "AIR COACH", 104-C

Fig. 213. Verville "Air Coach" model 104-C with 225 H.P. Wright J6 engine.

The sleek and trim-looking Verville "Air Coach" might be likened to most other cabin monoplanes of this period, but it did embody many worthwhile and original features that set it somewhat apart from the others. Among these many features was a rugged fuselage structure of a design that eliminated unsightly trussing through the seating and window area, and an interior coach-work that rivaled the tasteful beauty and spaciousness of the finest automobiles. As powered with the 7 cyl. Wright J6 engine of 225 h.p., the model 104-C was blessed with a reserve of performance and utility that was hard to beat; truly a design well worthy of the international reputation accumulated by Alfred Victor Verville, its designer.

The prototype version of the Verville "Air Coach" began taking shape in late 1927; as tested by "Billy" Brock at Ford Airport in Dearborn, Mich. a year later, it was powered with the 7 cyl. Warner "Scarab" engine of 110 h.p. Performance of this combination was quite adequate by average standards but somewhat lacking in the "punch" that would be expected by sportsman-pilots and busy business-men; as a consequence, the next model appeared soon after with the new 5 cyl. Wright J6 engine of 165 h.p. Introduced to the public at the annual Detroit Air Show for 1929, this Wright-powered combination demonstrated a performance that was greatly improved; this model was also available with the 165 h.p. Continental A-70. The inevitable power race amongst aircraft manufacturers, to gain and to hold a competitive advantage, was forced upon the Verville "Air Coach" also and consequently the 7 cyl. Wright J6 engine of 225 h.p. became the standard powerplant installation. Embodying all the latest aerodynamic principles, plus some that were entirely new, the model 104-C was light and easy on the controls and delightful in its graceful behavior aloft. The Verville "Air Coach" model 104-C received its type certificate number on 11-2-29 and was manufactured by the hand-picked crew at Verville Aircraft Co. of Detroit, Mich.; some 6 examples or more of this model were built in a production period that extended into 1931. B. F. Everitt was the president; R. S. Deering was V.P. and general manager; E. A. "Pete" Goff was sales director; Alfred V. Verville was secretary and the chief of design and engineering; and Wm. S. "Billy" Brock, world-famous aviator of this early period, was one of the several pilots engaged by the company for flight test and development.

Listed below are specifications and performance data for the Verville "Air Coach" model 104-C as powered with the 225 h.p. Wright J6 engine; length overall 28'9"; hite

Fig. 214. Coach styling of 104 was comparable to fine auto.

overall 8'0"; wing span 44'0"; wing chord 81"; total wing area 270 sq.ft.; airfoil "Clark Y"; wt. empty 2166; useful load 1234; payload with 76 gal. fuel was 561 lbs.; gross wt. 3400 lbs., max. speed 130; cruising speed 110; landing speed 50; climb 900 ft. first min. at sea level; climb in 10 min. was 7000 ft; ceiling 16,000 ft.; gas cap. 80 gal.; oil cap. 6.5 gal.; cruising range at 12 gal. per hour was 650 miles; price at the factory was $12,000., later reduced to $10,500. As a seaplane on Edo twin-float gear, following figures will apply; wt. empty 2450; useful load 1300; payload with 80 gal. fuel was 575 lbs.; gross wt. 3750 lbs.; performance was reduced in proportion.

The fuselage framework was built up of welded chrome-moly steel tubing in a pattern of trusses that eliminated unsightly and annoying frame members in seating and window area; faired to shape with wooden formers

and fairing strips, the completed framework was covered in fabric and metal side-panels. Among other unusual features incorporated into the fuselage structure were streamlined metal-framed sponsons that served as landing gear and wing strut attach points, also doubling in duty as compartments for storage of tool kit, battery, and other small articles. The cabin interior was tastefully upholstered in mohair fabric, and all windows were of shatter-proof glass; baggage compartment was to the rear of the cabin with an allowance for 50 lbs. The wing framework was built up of solid spruce spar beams that were routed out to an I-beam section, with truss-type wing ribs built up of spruce and plywood; ailerons were metal-framed, the leading edge was covered with aluminum alloy sheet, and the completed wing framework was covered in fabric. The two fuel tanks were mounted in the root end of each wing-half, one each side

Fig. 215. Verville "Air Coach" prototype of 1928 powered with 110 H.P. Warner engine.

of the fuselage, and the oil tank was mounted in the engine compartment. The interesting part about the oil tank on the 104-C was the incorporation of a simple and efficient oil temperature control device; a device that was nothing more than 9 open-end tubes that were built into the tank and submerged in the oil. By means of an air-scoop, air was drawn through these tubes thus lowering the temperature of the oil in the tank by drawing off the heat; by adjusting the scoop inlet, the oil temperature could be regulated. The outrigger type landing gear of 96 inch tread was fastened to the fuselage sponsons at its upper end and employed oil-draulic shock absorbing struts; wheels were 32x6 and Bendix brakes were standard equipment. For ease in ground maneuvering individual wheel brakes were operated by stirrups on the rudder pedals, and the tail-wheel was full swivel. The fabric covered tail-group was built up of welded chrome-moly steel tubing with the aft-end of the fuselage faired into the rudder; the fin was ground adjustable and the horizontal stabilizer was adjustable in flight. Navigation lights, cabin an instrument lights, metal propeller, and either

electric inertia-type or Heywood compressed-air engine starter, were also standard equipment. The next development in the Verville "Air Coach" was the Packard Diesel powered model 104-P, which will be discussed in the chapter for ATC # 316.

Listed below are Verville 104 entries that were gleaned from registration records:

X-151E;	Air Coach	(# 1)	Warner 110.
-506;	„	(# 2)	„
-303H;	„	(# 3)	Wright J6-5-165.
NC-303V;	„	(# 4)	„ J6-7-225.
NC-356V;	„	(# 5)	„ „
NC-70W;	„	(# 6)	Packard Diesel.
NC-477Y;	„	(# 7)	Wright J6-7-225.
NS-11;	„	(# 8)	
;	„	(# 9)	„ „
NS-3;	„	(# 10)	Wright J6-7-225.

Serial # 4 to # 10 eligible for this certificate; serial # 1, # 2, and # 3 were experimental prototypes, serial # 2 also had Wright J6-5-165; serial # 6 was model 104-P with 225 h.p. Packard Diesel; serial # 7 also on Group 2 approval numbered 2-306; registration number for serial # 9 unknown.

Fig. 216. For increased performance, second "Coach" was powered with 5 cyl. Wright J6 engine.

Fig. 217. Smooth lines and careful workmanship made the Verville "Air Coach" a stand-out among similar types.

A.T.C. #268
(11-4-29)
CURTISS-ROBERTSON "ROBIN", MODEL W

Fig. 218. "Robin" model W with 110 H.P. Warner engine; long nose was unusual but cute.

In the course of manufacturing several hundreds of aircraft, a manufacturer is surely bound to come up with a real rare example once in a while; the model W discussed here somewhat in a vague fashion is a real rare "Robin" indeed. From what is known of this version, the model W was also a 3 place high winged cabin monoplane typical to other "Robin" models, in most all respects, except for its powerplant installation which in this case was the 7 cyl. Warner "Scarab" engine of 110 h.p. A clear-cut description for character and other details of this Warner-powered model has never been available but from heresay and other accounts received we can safely assume that the nature of this rare version would be typical. Typical except for peculiarities of the extended nose-section which was necessary to mount the little Warner engine far enough out for proper balance around the point of C. G. Performance in most respects would be somewhat better than the OX-5 powered "Robin" but the "Model W" seemed to possess no other special merits that would warrant its production in any sizeable number.

The type certificate number for the

"Robin" model W was issued 11-4-29 for serial # 252; this one example was previously an OX-5 powered version and was modified to the specifications of ATC # 268 by the installation of the Warner 110 h.p. engine and other modifications necessary for this particular combination. Later specifications for ATC # 268 also allow a Warner "Scarab" engine of 125 h.p. Lest it be assumed that this was the only rare version of the popular "Robin", we hasten to add that there were several others of equal rarity. One version of the 3 place "Robin" was powered with the 5 cyl. Kinner K5 engine of 100 h.p. and was used for a time by the Kinner Airplane & Motor Co. as a test-bed for a further development of the K5 engine. One other version mounted the Curtiss "Crusader", a 6 cyl. inverted in-line aircooled engine of 120 h.p. that was flown in the "Robin" for many hours and showed good promise of being a successful and dependable powerplant; another version mounted the vee-type 8 cyl. water-cooled Hispano-Suiza (Hisso) model A engine of 150 h.p. The "Robin" model W was also built in one example that was modified for the Air Corps and delivered in late 1929; designated as the XC-10, it was in test as a light personnel transport. The XC-10 carried 3 at a gross weight of 2600 lbs., at a top speed of 112 m.p.h.; modifications to the fin area and its peculiar high-thrust engine mounting caused it to become rather ugly and somewhat out of proportion.

Listed below are specifications and performance data for the "Robin" model W as powered with the 110 h.p. Warner "Scarab" engine; length overall 25'10"; hite overall 7'10"; wing span 41'0; wing chord 72"; total wing area 224 sq.ft.; airfoil Curtiss C-72; wt. empty 1520 lbs.; useful load 780; payload with 50 gal. fuel was 275 lbs.; payload with 30 gal. fuel was 395 lbs.; gross wt. 2300 lbs.; max. speed 105; cruising speed 89; landing speed 45; climb 460 ft. first min. at sea level; ceiling 10,500 ft.; gas cap. normal 30 gal.; gas cap. max. 50 gal.; oil cap. 5 gal.; cruising range at 6.5 gal. per hour was 315-600 miles; no price was announced for this model.

The fuselage framework was built up of welded chrome-moly steel tubing, lightly faired to shape and covered with fabric. Interior arrangements would be typical to any other "Robin". The semi-cantilever wing framework was built up of solid spruce spar beams with "Alclad" aluminum alloy stamped-out wing ribs; the leading edge was covered with dural sheet and the completed framework was covered in fabric. Two fuel tanks were mounted in the wing, one each side of the fuselage. The split-axle landing gear of 94 inch tread used either spools of rubber shock-cord or oil-spring shock absorbing struts; wheels were 26x4 and no brakes were provided. A baggage compartment to the rear of the cabin was of 12.5 cu. ft. capacity for an allowance of 102 lbs. The fabric covered tail-group was built up of welded chrome-moly steel tubing and sheet steel formers; the fin was ground adjustable and the horizontal stabilizer was adjustable in flight. Other features provided as standard equipment were a steerable tail skid, a Curtiss-Reed metal propeller, and dual joy-stick controls. The next "Robin" development was the 4 place model 4-C as described in the chapter for ATC # 270 in this volume.

Listed below is the only known entry of the "Robin" model W as gleaned from various records:
NC-8376; Robin W (# 252) Warner 110.

NC-8376 was first registered as an OX-5 powered model B, later modified as model W and registered to Warren T. Jamieson of Edmond, Okla.; serial number for XC-10 unknown; a Robin (ser. # 384) was registered (NC-13H) in 1929 as model W but later modified to model J-1 with Wright R-540 engine; serial number for VH-UJE unknown.

A.T.C. #269
(11-6-29)
CURTISS "FLEDGLING", J-2

Fig. 219. Curtiss Fledgling model J-2 with 225 H.P. Wright J6 engine.

This next development in the popular 2 place open cockpit Curtiss "Fledgling" trainer was the model J-2, a progressive development in the series that was more or less typical to the standard U.S. Navy version designated the N2C-2. The model J-2 as shown here, was also basically typical to all other versions in the "Fledgling" series, except for its powerplant installation which in this case was the 7 cyl. Wright J6 engine of 225 h.p. With the increase in horsepower the model J-2 was somewhat faster and a bit more lively but otherwise it was also amiable and quite gentle; reports from various pilots who had a chance to fly this version would lead one to believe that it was by far the most popular craft in the whole series. With its increase in performance and consequent requirements for a slightly higher caliber of piloting technique, the J-2 was used mostly for training in the advanced phases of pilot instruction. The model J-2 was actually scarce in number at first, but it was easily possible to convert any existing "Fledgling" to the specifications called for under ATC

269. The model J-2 and all other "Fledgling" versions for that matter, were also available in the single-bay wing cellule typical of the "Fledgling Jr." (refer to chapter for ATC # 182 in U.S. CIVIL AIRCRAFT, Vol. 2). The "Fledgling Junior" with its single-bay wings of much less span offered a slight increase in top speed but apparently other advantages were lacking to some degree because it remained scarce in number.

The "Fledgling" model J-2 as powered with the Wright J6 engine of 225 h.p. received its type certificate number 11-6-29 and at least 2 or more examples of this model were manufactured by the Curtiss Aeroplane & Motor Co. at its Buffalo, New York division. The Curtiss company operated 3 plants at this time; one in Garden City, L.I., N.Y. and 2 in Buffalo. The Garden City facility was devoted entirely to design, development, and experimental work on new airplanes and the 2 plants at Buffalo were handling the quantity pro-

Fig. 220. Navy version of Fledgling J-2 was N2C-2.

duction; one plant was devoted to the manufacture of aircraft and the other plant manufactured the various Curtiss aircraft engines.

Listed below are specifications and performance data for the Curtiss "Fledgling" model J-2 as powered with the Wright J6 (R-760) engine of 225 h.p.; length overall 28'3"; hite overall 10'4"; wing span upper 39'1"; wing span lower 39'5"; wing chord both 60"; wing area upper 188 sq.ft.; wing area lower 177 sq.ft.; total wing area 365 sq.ft.; airfoil Curtiss C-72; wt. empty 2117 lbs.; useful load 888; crew wt. 380 lbs.; gross wt. 3005 lbs.; max. speed 115; cruising speed 98; landing speed 48; climb 960 ft. first min. at sea level; climb in 10 min. was 7350 ft.; ceiling 15,200 ft.; gas cap. normal 40 gal.; gas cap. max. 70 gal.; oil cap. 5 gal.; cruising range with 40 gal. fuel was 3.2 hours; range

Fig. 221. Curtiss Fledgling export model.

with 70 gal. fuel at 12.5 gal. per hour was 5.5 hours or 542 miles; baggage allowance was 48 lbs.; other dimensions characteristic of all "two-bay wing" versions include interplane gap at 68"; stagger 28.75" positive; incidence of upper & lower wings plus 2 deg.; dihedral of upper and lower wings was 1.5 deg.; in later models, a slight increase in performance was noticeable with the installation of the Wright J6 R-760 engine rated at 240 h.p.

The fuselage framework was arranged with two open cockpits in tandem; bucket seats were adjustable and had deep wells to fit a parachute pack. Wing walk-ways and convenient steps were provided for easy entry and cockpit coamings were well padded to avoid injury; a small baggage compartment was in the turtle-back section of the fuselage just behind the rear cockpit. Fuel tanks were mounted in the fuselage ahead of the front cockpit and oil supply tank was mounted on the fire-wall of the engine compartment. The rugged wing framework, because of its excessive span, was arranged as a two-bay wing cellule with two N-type wing struts on each side; the 4 ailerons were connected together in pairs and the pairs operated independently of each other by metal push-pull tubes. The vide tread landing gear was designed for

stability in landing and ease of ground maneuvering, also rugged enough to absorb abnormal abuse; wheel were 28x4 or 30x5 and wheel brakes were available. The fabric covered tail-group employed aerodynamic "balance horns" for both elevators and rudder to ease the control pressures; the fin was ground adjustable and the horizontal stabilizer was adjustable in flight. All mechanisms requiring frequent inspection or servicing were easily accessible through conveniently located removable panels; all points of wear were provided with lubrication fittings. It is suggested that reference be made to the chapters for ATC # 182 and # 191 of U.S. CIVIL AIRCRAFT, Vol. 2 for all other details of configuration and structure that are typical of all models in the "Fledgling" series.

Listed below are Curtiss "Fledgling" model J-2 entries as gleaned from registration records:
NC-463K; Fledgling J-2 (# B-67)
                                    Wright J6-7-225.
NR-274H;   „   (# B-102) „

Serial # B-67 and # B-102 were first as "Fledgling Junior" with Curtiss "Challenger" engine; several craft of the model J-2 type were exported to foreign countries.

Fig. 222. Model J-2 was probably best-liked of the Fledgling series.

Fig. 223. As modified to seat 4, the front passenger in the 4-C would share chummy quarters.

In an effort to make the popular "Robin' cabin monoplane into a still more useful carrier, Curtiss engineers dwelled on the possibilities of enlarging the capacity of this 3 place craft to seat 4 places, deciding first to modify an existing "Robin" model C to study the changes that would be necessary. From the outset, it was apparent that the front portion of the cabin, which normally seated only the pilot, would be rather chummy to say the least for the seating of two in this narrowed space. This led to a decision to increase the fuselage width by 4 inches in the later models of this 4-seated version. A small batch of 3 airplanes in the 4-seated version were built under a Group 2 approval as the model 4C-1 and a batch of at least 2 were built under the approval of ATC # 309 as the model 4C-1A. It is doubtful if any more than 6 or 7 examples of the 4-seated "Robin" were built in all and these actually were only a prelude to the new Curtiss "Sedan" series that came out shortly after.

The Curtiss-Robertson "Robin" model 4-C was basically typical in most all respects except for the changes in cabin arrangement to allow the seating of four passengers; modified from an existing model C, the original powerplant in this example was the 6 cyl. Curtiss "Challenger" engine of 170 h.p. which was later replaced with the improved version

of this engine rated at 185 h.p. From the specifications of the model 4-C it is apparent that only 420 lbs. would be available for the 3 passengers (not counting the pilot) when fueled to a maximum capacity of 50 gal.; this would only allow an average of 140 lbs. per passenger. With a minimim fuel load of 30 gal. the available payload would then rise to 540 lbs. which would allow the normal average of 165 lbs. for each passenger, with an allowance left over for 45 lbs. of baggage. In order to carry the 102 lbs. of baggage that the specification allowed, it is apparent that someone would have to be left behind. The type certificate number for the model 4-C was issued 11-6-29 and the modification of this airplane to the new specifications was believed to have been performed in the plant of the Curtiss-Robertson division at St. Louis (Anglum), Mo. Maj. Wm. B. Robertson was the president; and the aggressive and hard-working Ralph Shephard Damon was V.P. and manager of production.

Listed below are specifications and performance data for the "Robin" model 4-C as powered with the "Challenger" engine of 185 h.p.; length overall 25'1"; hite overall 8'0"; wing span 41'0"; wing chord 72"; wing dihedral 1.5 deg.; total wing area 224 sq.ft.; airfoil Curtiss C-72; wt. empty 1676 lbs.; useful load 924; payload with 30 gal. fuel was

540 lbs.; payload with 50 gal. fuel was 420 lbs.; gross wt. 2600 lbs.; max. speed 120; cruising speed .102; landing speed 47; climb 640 ft. first min. at sea level; climb to 5000 ft. was 8 min.; ceiling 12,700 ft.; gas cap. normal 30 gal.; gas cap. max. 50 gal.; oil cap. 5-6 gal.; cruising range min. fuel was 300 miles; cruising range max. fuel was 500 miles; tentative price at factory for model 4-C conversion was $7995.

Outside of cabin arrangement to allow the seating of 4 passengers, the model 4-C was typical in construction and aerodynamic arrangement to the "Robin" of the model C

and C-1 versions; for a detailed discussion of these models refer to the chapters for ATC # 69 in U.S. CIVIL AIRCRAFT, Vol. 1 and ATC # 143 in Vol. 2. Other details pertinent to the model 4-C were as follows; baggage capacity of 12.5 cu. ft. with allowance for 102 lbs.; wheel tread 96 in.; wheels 28x4 with Bendix brakes; steerable tail-wheel; Curtiss-Reed metal propeller; and speed-ring engine cowling.

Listed below is the only known example of the "Robin" model 4-C as gleaned from various records:

NC-8336; Robin 4-C (# 208) Challenger 185.

A.T.C. #271
(11-8-29)
BACH "AIR YACHT", 3-CT-9

Fig. 224. Bach "Air Yacht" model 3-CT-9, a medium-sized tri-motor of high performance.

The Bach tri-motored "Air Yacht" which was designed as a medium capacity transport capable of a very high performance, was an unusual airplane in many ways. Probably the most noticeable departure from the average lay-out was the mixed power combinations that were used; nearly all of the other tri-motors up to now were using identical engines of matched horsepower but Bach in their great variety of models, always had one big powerful brute in the nose-section in combination with two much smaller engines in the wing nacelles. Despite such a questionable set-up, the Bach "Air Yacht" was a craft with plenty of muscle and quite hard to beat in all-round performance. From each successive model it can be noticed that Bach was now tending to break away slightly from such odd engine combinations and the differential in horsepower was not nearly as great. Built at a time when most everyone else was making a switch to metal-framed fuselage structures, the Bach transport was purposely built of wood to soften and deaden the engine noises and help absorb operating vibrations; though not up-to-date with the general trend the wooden structured airplane still had considerable merit in its favor that worked out very well in practice and in principle.

Like most manufacturers at this time, Bach had several models in development and models currently in production were continually being up-graded even while on the line to keep pace with competition and increased demands by aircraft operators. In answer to these demands dictated by a growing trend, Bach developed the model 3-CT-9 which was a craft possessing a total of 900 h.p., the most powerful Bach tri-motor up to this time. This combination of thundering power translated into short and quick take-offs from the smaller fields, an excellent climb-out over surrounding obstacles, and a top speed well above the average for a craft of this particular type. In July of 1929 Waldo Waterman coaxed a Bach off with 1000 kilograms (2204 lbs.) aboard to a new altitude record of 20,820 ft. In a display of flashing speed and remarkable maneuverability, Waldo Waterman and Billy Brock, each flying a Bach "Air Yacht", finished 1-2 in the speed event for multi-motored aircraft over a 5 mile closed course of ten laps duration. This run-away was at the National Air Races for 1929 held at Cleveland, O.; the winner clocking an average of better than 136 m.p.h. and both polishing the pylons with the agility of a sport-plane. W. J. "Pat" Fleming later hoisted a 3-CT-9 off the ground with 4409 lbs. aboard and set a speed record of 144.60 m.p.h. for multi-motored craft in this category. The Bach was not often in competition but whenever they were, records usually fell to their onslaught.

The tri-motored Bach "Air Yacht" model 3-CT-9 was a high-winged cabin monoplane with ample seating for 10 in good comfort and practical elegance. Powered with one 9 cyl. Pratt & Whitney "Wasp" engine of 450 h.p. in the nose and two 7 cyl. Wright J6 (R-760) engines of 225 h.p. each in the wing nacelles, this was a combination with plenty of power reserve for a rather high performance. An example of this model was put into service on a west coast air-line to maintain faster schedules. Though quite sparse in actual number built, the Bach "Air Yacht" was a colorful ship with a personality that carved itself a memorable niche in the annals of airplane development. Beset by problems that were instigated by the economic depression, production schedules were cut drastically in the first part of 1930 and only a few Bach "tri-motors" were built during this shaky period; it was not too long after that Bach was forced to give up and call it quits. The type certificate number for the Bach "Air Yacht" model 3-CT-9, as powered with one 450 h.p. "Wasp" engine and two "Whirlwind Seven" engines of 225 h.p. each, was issued 11-8-29 and some 2 or more examples of this model were manufactured by the Bach Aircraft Co., Inc. at the Los Angeles Metropolitan Airport in Van Nuys, Calif.

Listed below are specifications and performance data for the Bach "Air Yacht" model 3-CT-9; length overall 36'10"; hite overall 9'9"; wing span 58'5"; wing chord at root 132"; wing chord at tip 96"; total wing area 490 sq.ft.; airfoil Clark Y of tapering depth percentage; wt. empty 5010 lbs.; useful load 2990; payload with 195 gal. fuel was 1525 lbs.; gross wt. 8000 lbs.; max. speed 162; cruising speed 136; landing speed 60; climb

1380 ft. first min. at sea level; ceiling 20,190 ft.; gas cap. 195 gal.; oil cap. 16 gal.; cruising range at 42 gal. per hour was 600 miles; price at the factory field was $39,500., lowered to $32,000. late in 1930.

The fuselage framework was built up of six wooden longeron members that were bolted together with steel gusset plates and fittings; the framework was cleverly arranged so that the cabin area was clear of all braces and there were no bracing obstructions in the doors or windows. The completed framework was covered with plywood veneer and an outer covering of fabric for surface smoothness and added torsional strength. The cabin was fitted with windows of shatter-proof glass and arranged with 8 comfortable passenger seats; the pilot and co-pilot were seated in a separate cockpit section up front with access to the main cabin. A baggage compartment of 50 cu. ft. capacity was to the rear of the main cabin with access from the inside or the outside. The semi-cantilever wing framework was built up of spruce and plywood box-type spar beams with plywood wing ribs that were routed out with holes for lightness then reinforced with spruce diagonals and cap-strips; the leading edges were covered with plywood and the completed framework was covered in fabric. The wing was braced to the fuselage by two parallel struts of heavy-walled chrome-moly steel tubes that were faired to an Eiffel 376 airfoil section for stability and added lift; the engine nacelles were built into a rigid truss framework with the undercarriage and the wing bracing struts. The fuel tanks were mounted in the root ends of each wing half and fuel to the engines flowed through a

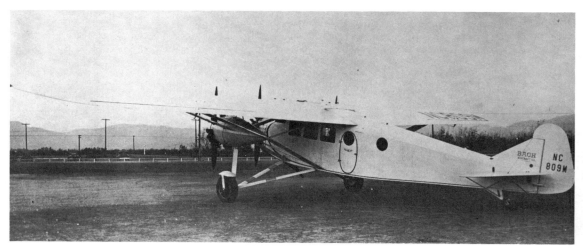

Fig. 225. 3-CT-9 was holder of several records for speed.

Fig. 226. Bach 3-CT-9 served several years with Union Air Lines.

selector valve in the pilot's cockpit. The plywood covered tail-group was built up in a wood framework similar to that of the wings; the vertical fin was built integral with the fuselage aft section and the horizontal stabilizer was adjustable in flight. The outrigger landing gear of 18 foot tread was fastened from the fuselage to the engine nacelles and used a combination of rubber-hydraulic shock absorbing struts; a dual tail-wheel was provided, main wheels were 36x8 and Bendix brakes were standard equipment. Metal propellers, inertia-type engine starters, and navigation lights were also standard equipment. The next development in the Bach "Air

Yacht" series was the plush model 3-CT-9S as described in the chapter for ATC # 299 in this volume.

Listed below are Bach "Air Yacht" model 3-CT-9 entries as gleaned from registration records:

NC-809M; Bach 3-CT-9 (# 19)
       Wasp 450 & 2 Wright R-760.
NC-511V;   „ (# 21)
       Wasp 450 & 2 Wright R-760.

Serial # 19 later as 7 place 3-CT-9 Special under Group 2 approval 2-175.

Fig. 227. Davis Model D-1-K with Kinner K5 engine; combination delivered pursuit-performance.

The petite and slender "Davis" parasol-winged monoplane was perhaps as good or just a little bit better than the average small airplane with 65 h.p. but with the added horsepower of a Kinner K5 engine, it really came to a sparkling life, as near to low-budget "pursuit performance" as one could get on 100 h.p. The take-off and climb-out characteristics of this combination were just short of terrific, coupled with a cruising and top speed that was certainly a match for the very best. Maneuverability and response was excellent and the type of flying capable with this eager craft was only limited by the pilot's mood or his ability; the pure flying pleasure built into and contained in this sport airplane was usually translated into an exhuberance that was a joy to experience or even to watch.

The Davis model D-1-K was a 2 place open cockpit sport-craft seating two in tandem and was basically similar to the model D-1 (refer to chapter for ATC # 256 in this volume) in most all respects except for the engine installation which in this case was the 5 cyl. Kinner K5 of 100 h.p. An approved type certificate for the model D-1-K was issued 11-12-29 and

some 10 or more examples of this model were manufactured by the Davis Aircraft Corp. at Richmond, Ind. Walter C. Davis, president of the company, was a former Captain in the Air Service and took special delight in demonstrating his company's wares; Lt. Lewis "Pat" Love, another U.S. Air Corps pilot, was the factory manager and chief pilot. Dwight Huntington was the chief engineer and about 30 craftsmen of various sorts were employed by the company at this time. During a visit to a small airport in 1953, a tattered and forlorn looking D-1-K was seen that had been hauled off to a back-lot where it was slowly rotting away; an attempt was made to buy the craft but the owner could not readily be found. Thinking it a down-right pity that so fine a ship should be relegated to such an end, you can imagine our pleasant surprise when we saw this very same ship flying again, as good as new, some 10 years later.

Listed below are specification and performance data for the Davis model D-1-K as powered with the 100 h.p. Kinner K5 engine; length overall 20'4"; hite overall 7'3"; wing span 30'2"; max. wing chord 63"; wing chord

Fig. 228. Parasol wing on D-1-K placed well for visibility.

mean 56″; total wing area 145 sq.ft.; airfoil Goettingen 387 mod. at max. thickness, tapering to Clark Y at root and tip; wt. empty 925 lbs.; useful load 536; payload with 25 gal. fuel was 197 lbs.; gross wt. 1461 lbs.; max. speed 127; cruising speed 105; landing speed 42; climb 1050 ft. first min. at sea level; climb in 10 min. was 8000 ft.; ceiling 14,000 ft.; gas cap. 25 gal.; oil cap. 2.5 gal.; cruising range at 6 gal. per hour was 410 miles; price at the factory was $4185., soon lowered to $3995.; later in 1930 this was lowered to $2995. and finally to $2295. in early 1932 as the cheapest Kinner-powered airplane available.

The fuselage framework was built up of welded chrome-moly and 1025 steel tubing, heavily faired to an oval shape with wooden formers and metal fairing strips then fabric covered. There was an entrance door on the left side for the front cockpit and there was a small baggage compartment with allowance for 27 lbs. in the turtle-back section behind the rear cockpit. Convenient steps were provided to either cockpit. The semi-cantilever wing framework was built up of laminated spruce spar beams and dural wing ribs built up of rounded channel sections; the leading edges were covered with dural sheet and the completed framework was covered in fabric. The center-section portion of the wing which contained the 25 gal. fuel tank, was a welded steel tube structure and fastened to the

fuselage by streamlined steel tube struts. The elevator and aileron controls were actuated by steel push-pull tubes and the rudder was actuated by stranded steel cable. The split-axle landing gear was of two long telescopic legs fastened to vees from the lower part of fuselage; tread was 78 inches, shock absorbers were spools of rubber shock-cord, wheels were 26x4 and brakes were available. Dual joy-stick controls were provided with either stick being quickly removable and rudder pedals could be quickly disconnected; bucket seats were provided with wells to fit a parachute pack. The fabric covered tail-group was built up of a welded steel tube frame and sheet steel former ribs; the fin was ground

Fig. 229. Walter C. Davis beams proudly from cockpit of Davis monoplane.

Fig. 230. "American Moth" lineage still apparent in Davis configuration.

adjustable and the horizontal stabilizer was adjustable in flight. Among the special equipment available for the Davis D-1-K were metal panel to cover front seat when not in use, a special streamlined windshield for the rear cockpit, wheel brakes, metal propeller, streamlined wheel fairings, and Heywood air-operated engine starter. The next development in the "Davis" parasol monoplane was the model D-1-66 with 85 h.p. LeBlond engine and as described in the chapter for ATC

317.

Listed below are Davis model D-1-K entries as gleaned from various records:

NC-77N;	Davis D-1-K	(# 501)	Kinner K5.
NC-842N;	„	(# 502)	„
NC-857N;	„	(# 503)	„
NC-858N;	„	(# 504)	„
NC-648N;	„	(# 505)	„
NC-649N;	„	(# 506)	„
NC-650N;	„	(# 507)	„
NC-158Y;	„	(# 508)	„
NC-157Y;	„	(# 509)	„
NC-151Y;	„	(# 510)	„
NC-156Y;	„	(# 511)	„

Serial # 501 was on Group 2 approval 2-148; registration number for serial # 502 and # 506 unverified; serial # 504 later as D-1-W on Group 2 approval 2-394; it is believed that D-1-K production went beyond serial # 511 but there is no proof.

Fig. 231. Performance and character of D-1-K ideal for the sportsman.

A.T.C. #273
(11-13-29)
ST. LOUIS "CARDINAL", C2-60

Fig. 232. St. Louis "Cardinal" model C2-60 with Le Blond 60 engine; carefully designed for the private-owner.

Added to the swelling list of light cabin monoplanes that were being offered in 1929, was the St. Louis "Cardinal". The pert "Cardinal" monoplane was not planned to be a radical departure from the normal in any way so it was quite typical to the average and the plan was only to offer a small ship of this type with a few extra features that were not usually found. As good references some 40 odd years of experience in developing and manufacturing of transportation vehicles and equipment were behind the development of the St. Louis "Cardinal"; the St. Louis Car Co., parent organization of the St. Louis Aircraft Corp., was one of the enlisted manufacturers that built war-time aircraft for the U.S. government during the pioneering days of World War 1. Anxious to re-enter aircraft manufacture again, the "Cardinal" monoplane was developed and tested during the hustle and bustle of 1928 and proudly exhibited to the flying public at the Detroit Air Show for 1929; the company followed up the spurt of interest with a short period of advertising and bally-hoo but the aspiring craft was actually little known and very rarely seen in many parts of the country. Because of this, there is very little lore left behind by this little airplane but at least one has been rebuilt to fly again in recent years.

The St. Louis "Cardinal" model C2-60 as

shown here, was a high winged cabin mono-plane seating two side by side with good visibility and ample comfort; the powerplant for this particular model was the 5 cyl. Le-Blond 5-D engine of 65 h.p. and performance was a good average for a ship of this type. Planned early enough to be the basis for several different models, the "Cardinal" structure was beefy enough to withstand the stresses of higher-powered engines. Other extra features not usually found in a small ship of this class, were wheel brakes, a swiveling tail-skid, oil-draulic shock absorbing struts in the landing gear, shatter-proof glass throughout, and interior appointments that were some-what better than the average. Flight charac-teristics and general behavior were not par-ticularly outstanding but always described as pretty good, so we can assume that the "Cardinal" was just about normal in every respect. The "Cardinal" C2-60 was first certi-ficated on a Group 2 approval numbered 2-92 (issued 7-15-29) for serial # 103 and up, with no dual controls, but this approval was later superseded by ATC # 273. The type certifi-cate number for the "Cardinal" model C2-60, as powered with the 65 h.p. LeBlond engine, was issued 11-13-29 and some 10 or more examples of this model were manufactured by the St. Louis Aircraft Corp. at St. Louis, Mo.; a subsidiary of the St. Louis Car Co. The car company and the aircraft plant were both at the same address so apparently the "Cardinal" was manufactured in a corner of the street-car factory.

Listed below are specifications and performance data for the "Cardinal" model C2-60 as powered with the LeBlond 5-D engine; length overall 20'7"; hite overall 7'0"; wing span 32'4"; wing chord 60"; total wing area 162 sq.ft.; airfoil Clark Y; wt. empty 929 lbs.; useful load 521; payload with 25 gal. fuel was 196 lbs.; gross wt. 1450 lbs.; max. speed 100; cruising speed 85; landing speed 36; climb 570 ft. first min. at sea level; ceiling 9750 ft.; gas cap. 25 gal.; oil cap. 1.5 gal.; cruising range at 4.5 gal. per hour was 425 miles; price at the factory was first quoted at $2950., soon raised to $3250. but reduced to $2950. in May 1930.

The fuselage framework was built up of welded chrome-moly and 1025 steel tubing in a rigid Warren truss, faired to shape with wooden formers and fairing strips then fabric covered. The cabin interior was neatly appointed in long-wearing fabrics and the car-type seat had removable cushions to accommodate parachute packs; a baggage compartment of 4 cu. ft. capacity with allowance for 14 lbs. was located in back of the seat. Windows were of shatter-proof glass with a liberal amount of area for ample visibility; a sky-light was provided in the roof for vision overhead and window side-panels offered vision in the wheel-path for the pilot. The wing framework was built up of laminated spruce spar beams that were routed out for lightness, with spruce and basswood truss-type wing ribs; the leading edge was covered to preserve the airfoil form and the completed framework was covered with fabric.

The two gravity-feed fuel tanks were mounted in the wing flanking the fuselage; the wing was braced to the fuselage by a pair of parallel steel tube struts. The outrigger landing gear of 75 in. tread was incorporated with the wing bracing struts into a rigid truss; shock absorbing struts were oil-draulic, wheels were 26x4 and wheel brakes were standard equipment. The tail-skid was mounted on the very end of the fuselage and swiveled to allow unhampered ground maneuvering. The fabric covered tail-group was built up of welded steel tubing; the fin was ground adjustable and the horizontal stabilizer was adjustable in flight. The next development in the "Cardinal" monoplane was the Kinner powered "Super Cardinal" model C2-110 as described in the chapter for ATC # 277 in this volume.

Listed below are "Cardinal" model C2-60 entries as gleaned from registration records:

X-422;	Model C2-60	(# 101)	Le Blond 60.
C-530E;	„	(# 102)	„
NC-31H;	„	(# 103)	„
NC-360K;	„	(# 104)	„
NC-903K;	„	(# 105)	„
NC-951K;	„	(# 106)	„
NC-989K;	„	(# 107)	„
NC-990K;	„	(# 108)	„
;	„	(# 109)	„
;	„	(# 110)	„

Serial # 101 and # 102 were both experimental prototypes and not eligible for this approval; serial # 103 later converted to model C2-90; serial # 105 later had 80 h.p. Genet engine; serial # 106 later converted to model C2-110; serial # 107 also listed as C2-90; registration numbers for serial # 109-110 unknown.

Fig. 233. First production version of "Cardinal" C2-60 harbored several improvements.

Fig. 234. Original prototype of the St. Louis "Cardinal."

Fig. 235. Rogers "Sea Eagle" with 225 H.P. Wright J6 engine; sporty cockpit afforded open-air fun.

A general use of the nation's natural and ready-made landing places, the rivers, lakes, bays, and sea-coast by airplanes, still remained to be fully achieved; though the potential advantages to be gained from off-water operations were barely scratched, an interest in the use of the seaplane, flying boat, and amphibian for commercial and private uses continued to grow. Several years back the seaplane or flying boat was quite popular in relation because good airports for land-craft were few and far between but when airports became more plentiful the popularity of seaplanes began to fall off. In 1927 interest began to pick up again and several new designs sprang forth to help revive the art of flying off water; by the end of 1929 some 15 different types were offered as flying boats and amphibians, and some 45 land-craft were convertible to operate as seaplanes on pontoons. Without a doubt the greatest interest for the seaplane type came from the private-owner who would normally use his craft for pleasure-hops somewhat in the manner of speed-boat riding or perhaps to fly to some resort or hunting grounds; many good airplanes were built for just such a purpose and the Rogers "Sea Eagle" posed to be one of them. As pictured here the Rogers "Sea Eagle" was laid out with an open cockpit

in the forward part of the hull and its other old-fashioned lines seemed to indicate that it was meant to be sort of a flying speed-boat for the private-owner, just for the sport of it, and not particularly planned for any commercial use. It is surprising that more people didn't take to flying off water — it was great fun; perhaps it was the excitement of boating and the thrill of flying brought together that made it such a memorable experience.

Harry Rogers, probably the most experienced "flying boat" pilot in the country, designed the "Sea Eagle" primarily for the sportsman-pilot or for the small operator who would specialize in local charter flights or joy-hop passengers on a nice week-end. Patterned after a proven design, the Rogers flying-boat demonstrated excellent flight characteristics, its behavior in the water was very good and though performance was not exceptional, it was adequate for the purpose. Operating the Rogers Air Lines at Garden City, Long Island and also in Miami, Florida, Harry Rogers flew Curtiss "Sea Gull" flying boats for many years and operated a flying service that specialized in over-water flights; his regard for the "Sea Gull" must have been rather high because the Rogers "Sea Eagle" was very much like the Curtiss "boat" only in more modern dress. The prototype example of the "Sea Eagle" (as X-9735) was powered with the 8 cyl. Hispano-Suiza (Hisso) engine but a quest for better performance suggested the need for more horsepower. Powered later with the 7 cyl. Wright J6 engine of 225 h.p. the "Sea Eagle" model RBX could take off at dead-calm in 16 sec. with a full gross load and climb-out was greatly improved. The RBX was a biplane of the "flying boat" type with a large open cockpit up forward that had ample seating for five; despite its old-fashioned and drafty appearance the robust hull was covered with a metal skin and wing frames were of the most modern construction with the efficient Curtiss C-72 airfoil. The engine nacelle was mounted midway between the wings in "pusher" fashion, a configuration that offered many advantages and was found to be the most practical for a craft of this type. Harry Rogers used the first "Sea Eagle" in his own service and as far as can be determined, this was the only example ever built. The type certificate number for the Rogers "Sea Eagle" model RBX as powered with the 225 h.p. Wright J6 (R-760) engine was issued 11-14-29 and it was manufactured by Rogers Aeronautical Mfg. Co., Inc. at Roosevelt Field in Long Island, N. Y. From what little has been written about this craft we can assume that Harry Rogers was head of operations and the chief of design.

Listed below are specifications and performance data for the Rogers "Sea Eagle" model RBX as powered with the 225 h.p. Wright J6 engine; length overall 32'0"; hite overall 10'10"; wing span upper 40'0"; wing span lower 31'0"; wing chord upper 72"; wing chord lower 72"; wing area upper 222 sq.ft.: wing area lower 157 sq.ft.; total wing area 379 sq.ft.; airfoil Curtiss C-72; wt. empty 2396 lbs.; useful load 1034; payload with 50 gal. fuel was 520 lbs.; gross wt. 3430 lbs.; max.

speed 105; cruising speed 85; landing speed 43; climb 740 ft. first min. at sea level; ser. ceiling 12,000 ft.; gas cap. 50 gal.; oil cap. 7 gal.; cruising range at 14 gal. per hour was 270 miles; price at the factory was $12,500.

The simple hull framework was built up of ash and spruce members covered with "Alclad" sheet of varying gauges; the outer metal skin was screwed to the frame. The forward deck was covered with wooden planking and the open cockpit normally seated the pilot and 3 passengers. The drafty cockpit had a windshield up front for some protection and 2 narrow walk-ways on upper edges of the hull for entry to the seats. The wing framework was built up of solid spruce spars that were routed to an I-beam section, with stamped-out duralumin wing ribs; the leading edges were covered with "dural" sheet and the completed framework was covered in fabric. Interplane struts were of streamlined steel tubing and all interplane bracing was of stranded steel cable. Ailerons in the upper wing were actuated by push-pull struts and all other flight and engine controls were operated by stranded steel cable. The gravity-feed fuel tank was mounted in the center-section panel of the upper wing and the oil tank was mounted in the engine nacelle. The fabric covered tail-group was built up of welded steel tubing; the fin was ground adjustable and the horizontal stabilizer was adjustable in flight. Mooring gear was stored in the forward section of the hull and there was no provision for baggage. A metal propeller, navigation lights, hand-crank inertia-type engine starter, life-preserver seat cushions, and mooring gear were standard equipment.

Listed below is the only known entry of the Rogers "Sea Eagle" model RBX:
NC-9735; Sea Eagle RBX (# 1) Wright R-760

Fig. 236. Monocoach model 275 with 225 H.P. Wright J6 engine.

With some 14 or more examples of the Wright J5 powered "Monocoach" 201-type already built and in service, Mono Aircraft hastened to offer a new and slightly improved version called the "Model 275". The "Monocoach" model 275 was in fact more or less typical of the earlier model but now powered with the new 7 cyl. Wright J6 (R-760) engine of 225 h.p.; with this lighter and slightly more powerful engine installation, the new "Coach" was some 36 lbs. lighter when empty, carried 44 more lbs. in useful load and operated at a gross load some 8 lbs. heavier. This all translated into just a little more utility and a slightly noticeable increase in all-round performance. Typical of most light and medium sized airplanes of this day, the model 275 was designed to operate from hay-field airports and its short field performance with a full gross load was impressive indeed; aloft, it was still able to step out in good speed with an amiable nature and easily predictable handling characteristics. Never a harsh word has ever been said about the "Monocoach" and it certainly was a worthy addition to the star-studded "Mono" line as an outstanding value. While production lines were busy sending off the "Monocoupe 113", the "Mono-

prep 218", the "Monocoach 275", and an occasional "Monosport", several exciting things were happening on the side-lines. The new "Monocoupe 90" was being groomed to take place of the "113" and the "Monosport" was redesigned considerably to make into the fabulous "Monocoupe 110"; great events in Mono Aircraft history were just a short ways off.

The "Monocoach 275" as pictured here, was a high winged cabin monoplane that seated 4 with ample room and good comfort; designed as a family-type airplane or for air-taxi work in business, the "275" had inherent characteristics that were conducive to reliability in operation, the ability to show a fair profit, and a friendly nature that stimulated pride of ownership. As one owner put it; "it may not be the very best airplane of this type but it surely is one of the very best". One example of the model 275 was still flying in Boise, Idaho as late as 1960; the owner preferred it over new production craft for short field high altitude work. Several of the earlier "Monocoach 201" (ATC # 201) were later modified into the model 275 by installation of the 7 cyl. Wright J6 engine.

Fig. 237. Rugged structure and pleasant nature made "275" ideal family-type plane.

The type certificate number for the "Monocoach 275" was issued 11-13-29 and some 6 or more examples of this model were manufactured by the Mono Aircraft Corp. at Moline, Ill. That they were not built and sold in greater number is only an unfortunate circumstance caused by the "depression" of this period which had crippling effects on airplane sales, regardless of their caliber. The "Monocoach 275", slightly improved and reduced in price, was still being offered in 1931 but there were no buyers.

Listed below are specifications and performance data for the "Monocoach" model 275 as powered with the 225 h.p. Wright J6 engine; length overall 26'9"; hite overall 8'7"; wing span 39'0"; wing chord 75"; total wing area 230 sq.ft.; airfoil USA-35B; wt. empty 1883 lbs.; useful load 1217; payload with 63 gal. fuel was 626 lbs.; gross wt. 3100 lbs.; max. speed 130; cruising speed 112; landing speed 50; climb 900 ft. first min. at sea level; ceiling 18,000 ft.; gas cap. 63 gal.; oil cap. 6 gal.; cruising range at 12.5 gal. per hour was 530 miles; price at factory field was $8250.

The fuselage framework was built up of welded 4130 and 1025 steel tubing in a rigid truss form; the frame was faired to shape with wooden formers and fairing strips then fabric covered. The front seats were of the bucket-type with folding seat-backs and the rear seat was of the bench type; there was a large baggage compartment with allowance for 116 lbs. behind the rear seat and all windows were of shatter-proof glass. Entry was gained to the cabin with a convenient step and a large door on each side; the cabin walls were sound-proofed and insulated and the interior was tastefully upholstered in mohair fabrics. There was a large sky-light in the pilots area for vision upwards and dual joystick controls were provided; windows in the door panels slid up and down for ventilation and cabin heat was provided from the engine-exhaust system. The semi-cantilever wing framework, in 2 halves, was built up of solid spruce spars that were routed to an I-beam section, with wing ribs of spruce and basswood in truss-type form; the leading edges were covered with dural sheet and the completed framework was covered in fabric. The 2 fuel tanks were mounted in the root end of each wing half flanking the fuselage and the wing was braced by two parallel steel tube struts of streamlined section on each side; the landing gear was built into a sturdy truss with the wing-brace struts. The wide tread landing gear was of the out-rigger type and employed spring-draulic shock absorbing struts; wheels were 30x5 with Bendix brakes and a full-swivel tail wheel was provided for ease in ground maneuvering. The fabric covered tail-group was built up of welded chrome-moly and 1025 steel tubing; the fin was ground adjustable and the horizontal stabilizer was adjustable in flight. Optional color schemes were of course available for the "Monocoach" but most all examples were a shiny black fuselage with bright orange-yellow wings. A metal propeller, hand-crank inertia-type engine starter, and navigation lights were standard equipment. The next "Mono" development was the Lambert-powered "Monocoupe 90" as discussed in the chapter for ATC # 306.

Listed below are "Monocoach 275" entries as gleaned from registration records:
NC-100K; Monocoach 275 (# 5011)

			Wright R-760.
NC-114K;	„	(# 5012)	„
NC-134K;	„	(# 5025)	„
NC-135K;	„	(# 5026)	„
NC-150K;	„	(# 5027)	„
NC-164K;	„	(# 5028)	„
NS-22;	„	(# 5029)	„
NS-33;	„	(# 5030)	„

Serial # 5011-5012 were first as model 201; serial # 5029-5030 were with Dept. of Commerce.

Fig. 238. Parks P-2A with 165 H.P. Wright J6 engine; ship shown first as model P-2.

Born as a class-room project at the Parks Air College, the Parks-built airplane first came out as the Curtiss OX-5 powered model P-1 and shortly thereafter in an Axelson powered version of considerable improvement known as the model P-2. Several of the first airplanes were actually built by Parks students at the school, where they followed the pattern of development from the "pile of steel tubing" stage to a completed airplane with actual tests on the flight line. Satisfied their craft was a good airplane and was sure to find a market amongst the thousands of people that were buying airplanes, Parks launched an ambitious project that led to the manufacture of the first two versions in some 50 or more examples. The Axelson powered model P-2 was selling comparatively slow and Parks was not entirely happy with the acceptance shown this combination so the P-2A was introduced; the P-2A was but a progressive development that was powered with the 5 cyl. Wright J6 engine, an engine which was riding a crest-wave of popularity about this time anyhow. By now a subsidiary of the Detroit Aircraft Corp., Parks' destiny was naturally

dictated by the parent firm so when the depression-ridden aircraft market of 1930 forced everyone in the industry to take economy measures, the manufacture of "Parks" airplanes was transferred to a plant in the famous motor-city. A further belt tightening in 1931 forced additional measures within the corporation so it was decided to merge Parks aircraft manufacture with that of Ryan Aircraft; thus the "Parks" lost its former identity and became known as the "Ryan Speedster".

The Parks model P-2A was also a 3 place open cockpit biplane of the popular general-purpose type and was quite typical to the earlier model P-2 except for its powerplant installation which was the 5 cyl. Wright J6 (R-540) engine of 165 h.p. and other modifications necessary for this combination. The added horsepower brought about an increase in performance but flight characteristics and general behavior remained more or less the same; the P-2A was not outstanding but a good average for a craft of this type. To present a more attractive and appealing

Fig. 239. Parks P-2A ideal for business or sport; later called Ryan "Speedster."

appearance the new Parks P-2A (Ryan "Speedster") was offered in a variety of gay colors. One pleasing combination was a deep maroon fuselage with gold-colored wings and an accenting body stripe of gold; another combination was a bright red fuselage with cream-colored wings and a cream body stripe. Other color combinations were optional upon request. The type certificate number for the Parks model P-2A was issued 11-14-29 and some 10 or more examples of this model were manufactured by the Parks Aircraft Corp. at E. St. Louis, Ill., or at Detroit, Mich. Ray W. Brown, formerly with the Travel Air Co., was manager of sales and Pat Murphy was in charge of sales promotion for all eleven units of the Detroit Aircraft Corp. In June of 1932 manufacture had already ceased and all "Parks" designs were sold to Dean Hammond of Ypsilanti, Mich. who altered the design almost beyond recognition and introduced it as the "Hammond" 100.

Listed below are specifications and performance data for the Parks model P-2A (Ryan "Speedster") as powered with the 165 h.p. Wright J6 engine; length overall 23'0"; hite overall 9'3"; wing span upper 30'0"; wing span lower 28'8"; wing chord both 63"; wing area upper 154 sq.ft.; wing area lower 131 sq.ft.; total wing area 285 sq.ft.; airfoil "Aeromarine" 2-A; wt. empty 1483 lbs.; useful load 897; payload with 52 gal. fuel was 372 lbs.; gross wt. 2380 lbs.; max. speed 120; cruising speed 100; landing speed 45; climb 800 ft. first min. at sea level; ceiling 14,000 ft.; gas cap. 52 gal.; oil cap. 6 gal.; cruising range at 9 gal. per hour was 500 miles; price at the factory field was $6350., lowered to $6285. in 1931.

The fuselage framework was built up of welded chrome-moly steel tubing deeply faired to shape with wooden fairing strips and fabric covered. The cockpits were deep and well protected with a door on the left side providing entry to the front cockpit. The fuel tank was mounted high in the fuselage just ahead of the front cockpit and a baggage compartment of 1.3 cu. ft. capacity with allowance for 30 lbs. was in back of the rear seat. The wing framework was built up of solid spruce spars that were routed to an I-beam section, with spruce and plywood truss-type wing ribs; ailerons of wooden construction were on the lower wings and the completed wing panels were covered with fabric. The split-axle landing gear of 72 inch tread used rubber shock-cord in tension to absorb shocks and these "shock struts" were encased in streamlined metal cuffs; some of the latest models used "oleo" struts in the landing gear. Wheels were either 28x4 or 30x5 and Bendix brakes were standard equipment. The fabric covered tail-group was built up of welded chrome-moly steel tubing; the fin was ground adjustable and the horizontal stabilizer was adjustable in flight. A metal propeller and wiring for lights was also standard equipment; navigation lights, and a hand-crank inertia-type engine starter were optional equipment. No great volume of

Fig. 240. Parks P-2A at antique fly-in, 1960.

production was ever attained in the "Parks" series but at least 20 in various models were still flying some 10 years later and at least one was still flying in the sixties.

Listed below are Parks models P-2A entries as gleaned from registration records:

NC-902K;	Parks P-2A	(# 2972)	Wright R-540.
NC-965K;	„	(# 2981)	„
NC-967K;	„	(# 2983)	„
NC-968K;	„	(# 2984)	„
NC-502N;	„	(# 2991)	„
NC-499H;	„	(# 101)	„
NC-480M;	„	(# 102)	„
NC-8487;	„	(# 103)	„
NC-8490;	„	(# 106)	„
NC-8491;	„	(# 107)	„
NC-8492;	„	(# 108)	„

Serial # 2972 and # 2981 first as model P-2; serial # 2972 - 2981 - 2983 - 2984 - 2991 were Parks Aircraft numbers serial # 101 - 102 - 103 -106 - 107 - 108 were Detroit Aircraft numbers; registration number for serial # 107 - 108 unverified.

Fig. 241. Parks P-2A spends its 31st birthday at 1960 Fly-in.

ST. LOUIS "SUPER CARDINAL", C2-110

Fig. 242. St. Louis "Super Cardinal" model C2-110 with Kinner K5 engine.

Introduced about a month or so earlier in versions powered with 65 h.p. and 90 h.p., the model C2-110 was a sportier development of the "Cardinal" monoplane for those of more hankering and better means that wanted near-to "pursuit" plane performance in a light cabin airplane; hence the label "Super Cardinal". Though typical to the other two models in the series, the "Super Cardinal" was designed to offer that extra bit of performance so attractive to the play-boy pilot or the busy business-man that had frequent calls to make in areas far off the beaten path. With its ability to take-off quite short and climb out at a steep rate, the C2-110 no doubt operated admirably out of small pasture-airports that were still the back-bone of American aviation activities during this period. Powered with the 5 cyl. Kinner K5 engine of 100 h.p. the sprightly "Super Cardinal" was the pride and joy of the St. Louis Aircraft Corp. but its introduction into troubled times stifled its full potential and production never reached any great number. It is cold fact that of the 4 versions built in the "Cardi-

nal" series, not any one of them enjoyed a worthwhile measure of success; total production in two years time was about 21 airplanes. Like several other long-established manufacturers in other fields of business, the St. Louis Car Co., parent organization of the St. Louis Aircraft Corp., was enamored by the possibilities of a promising future in aircraft manufacture but the rosy bubble had soon bursted and the company was forced to call it quits after some two years in the business.

The stubby "Super Cardinal" was a light high winged monoplane that seated two comfortably in side by side arrangement; appointments were handy and in good taste and several extra features were provided to offer utility and on-the-line operation with a minimum of fuss. The Kinner K5 engine of 100 h.p. offered power reserve that translated into above-average performance and probably instigated a very playful nature; no complaints have ever been lodged against any of the "Cardinals", so we can assume that the

C2-110 was probably a good average for a high performance craft of this type. Simple of line and of good aerodynamic arrangement the "Cardinal" harbored no fluke theories nor boasted of any razzle-dazzle that would stand it apart; a fair price for what it had to offer was calculated to be reason enough for its introduction to the market. The type certificate number for the "Super Cardinal" model C2-110 was issued 12-4-29 and some 6 or more examples of this model were manufactured by the St. Louis Aircraft Corp. at St. Louis, Mo.; a subsidiary of the St. Louis Car Co., manufacturers of street-cars. It is odd coincidence that of the few vehicle and custom carriage factories that turned to aircraft manufacture during this period, not one of them enjoyed any measure of success.

Listed below are specifications and performance data for the St. Louis "Super Cardinal" model C2-110 as powered with the Kinner K5 engine; length overall 20'6"; hite overall 7'0"; wing span 32'4"; wing chord 60"; total wing area 162 sq.ft.; airfoil Clark Y; wt. empty 1006 lbs.; useful load 557; payload with 30 gal. fuel was 190 lbs.; gross wt. 1563 lbs.; max. speed 125; cruising speed 107; landing speed 42; climb 1000 ft. first min. at sea level; ceiling 15,000 ft.; gas cap. 30 gal.; oil cap. 3 gal.; cruising range at 6.5 gal. per hour was 475 miles; price at the factory was $4250., later reduced to $3985.

The construction details and general arrangement of the model C2-110 were typical to that of the model C2-60 as described in the chapter for ATC # 273 in this volume. The baggage compartment on the C2-110 was of 4 cu. ft. capacity with an allowance for 14 lbs. with a full gross load. The wing bracing struts on the model C2-60 were round steel tubes that were encased in balsa-wood fairings which were shaped to a streamlined section; the wing bracing struts on the C2-90 and C2-110 were steel tubes of streamlined section. The later models of the "Cardinal" had a small Micarta tail-wheel incorporated into the tail skid for operation on hard-surface runways. A Hartzell or Supreme wooden propeller was standard equipment; a metal propeller, Heywood air-operated engine starter, and navigation lights were offered as optional equipment. The C2-110 designation for the "Super Cardinal" was a misnomer that was to have resulted from a quick change of heart in powerplants; as originally planned the C2-110 was laid out for the 7 cyl. Warner "Scarab" engine of 110 h.p. but undetermined circumstances caused a switch to the Kinner K5 of 100 h.p. and the alloted designation remained unchanged. To complete the cycle of unrelated designations, a 1931 model of the "Cardinal" finally was powered with the 110 h.p. Warner engine and was labeled the model C2-100! From records that are available this model appears to be the last "Cardinal" built.

Fig. 243. High performance behavior of C2-110 was ideal for the sportsman.

Listed below are "Super Cardinal" model C2-110 entries as gleaned from registration records:

NC-951K;	Model C2-110	(# 106)	Kinner K5.	
NC-528N;	„	(# 112)	„	
NC-561N;	„	(# 117)	„	
NC-710M;	„	(# 118)	„	
NC-719M;	„	(# 119)	„	
NC-587N;	„	(# 120)	„	

Serial # 106 first as model C2-60; serial # 121 was model C2-100 (X-12319) with 110 h.p. Warner engine in 1931 in test with St. Louis Aircraft Corp.

Fig. 244 Pride of the "Cardinal" series, C2-110 remained scarce.

Fig. 245. Popular Travel Air model 10-D with 225 H.P. Wright J6 engine.

To round out their offerings in the cabin monoplane series, Travel Air developed and introduced the "Model 10", a medium-sized craft of typical form that was introduced about a year too late. With the two larger cabin monoplanes, the 6000-B and the A-6000-A, filling the needs for slightly more capacity, the "Ten" was especially designed as a family-type airplane for the seating of four. With its fairly high performance and effortless ability to operate from the smaller airports, the model 10 was just the right size for many business houses too or for the flying service operator running a stand-by air-taxi and charter service. It is apparent that Travel Air had planned the 10-series for a broad scale of needs and possible uses because the spread of horsepower was to be from 185 h.p. on the low side to 300 h.p. on the high side. A model 10-B was actually tested with the 300 h.p. Wright J6 engine and one version was to have mounted the 185 h.p. Curtiss "Challenger" engine but because of the way the market was shaping up, the "Ten" offerings were narrowed down to the 10-D version and other developments were cancelled. This is a lesser known model in the "Travel Air" monoplane series but its popularity in spite of the sagging market was good and firm; had the 10-D been introduced somewhat earlier it is quite likely that a good

many more would have been seen around the so-called "bush country" of the nation. Though built in regrettably small number, its usefulness promoted life and extended service beyond the normal and some were still flying in the late fifties.

The handsome "Travel Air" model 10-D was a high winged cabin monoplane of buxom and hardy proportion with ample seating and comfort for four; basically typical of other familiar models in the cabin monoplane series, only in scaled-down fashion, the model 10-D did have several innovations that set it somewhat apart. Foremost in the change of form was the bulge-type cabin enclosure in the forward portion for better visibility and other numerous changes not easily seen that promoted better operation and easier servicing. Powered with the 7 cyl. Wright J6 engine of 225 h.p., performance was very good in a broad range with power reserve left over to help out in the tight places. Its lasting popularity was a good indication of its compatible nature and its basic arrangement led it to inherit all of the fine qualities that made the "Travel Air" monoplane such a great favorite. From a casual comparison of the two, it is easy to see that the model 10 laid the ground-work for the model 15 Curtiss-Wright "Sedan" that was developed a year or so later.

The type certificate number for the model 10-D as powered with the Wright R-760 engine was issued 12-2-29 and some 11 examples of this model were manufactured by the Travel Air Co. at Wichita, Kansas.

Listed below are specifications and performance data for the "Travel Air" model 10-D monoplane as powered with the 225 h.p. Wright J6 engine; length overall 26'10"; hite overall 8'8"; wing span 43'6"; wing chord 74"; total wing area 240 sq.ft.; airfoil Goettingen 593; wt. empty 2130 lbs.; useful load 1270; payload with 70 gal. fuel was 640 lbs.; gross wt. 3400 lbs.; max. speed 126; cruising speed 106; landing speed 55; climb 675 ft. first min. at sea level; ceiling 13,000 ft.; gas cap. 70 gal.; oil cap. 6 gal.; cruising range at 12.5 gal. per hour was 550 miles; price at the factory field first quoted as $11,250. lowered to $8495. about mid-year 1930.

The fuselage framework was built up of welded chrome-moly steel tubing, heavily faired to shape with metal formers and wooden fairing strips then fabric covered. A large door and convenient step on each side provided easy entry to the interior which was tastefully upholstered in long-wearing fabrics; the front seats were individual bucket type and the rear seat was of the bench type. Ample

window area provided good visibility and there were provisions for ventilation and cabin heat. A good-sized baggage compartment with allowance for 125 lbs. was located behind the rear seat. The wing framework was built up of solid spruce spar beams with spruce and plywood truss-type wing ribs; the leading edge was covered with dural sheet and the completed framework was covered with fabric. Wing bracing struts were large-diameter steel tubes that were encased in balsawood fairings which were shaped to an airfoil section. Fuel tanks were mounted in the root end of each wing half flanking the fuselage and ailerons were of the "Friese" balanced-hinge type. The wide tread landing gear, built into a strong truss with the wing bracing struts, was of the out-rigger type with oleo-spring shock absorbing struts; wheels were 30x5 and Bendix brakes were standard equipment. The fabric covered tail-group was built up of welded chrome-moly steel tubing and sheet steel forming ribs; the fin was ground adjustable and the horizontal stabilizer was adjustable in flight. A 14x3 tail wheel was mounted on aft portion of the fuselage to swivel easily for good ground maneuvering. A metal propeller, navigation lights, Bendix wheel brakes, were standard equipment. A hand-crank inertia-type engine starter and cabin heater was optional. The

Fig. 246. Early version was model 10-B with 300 H.P. Wright J6 engine.

next "Travel Air" development was the model 4-P biplane as described in the chapter for ATC # 280 in this volume; the next "Travel Air" monoplane development was the model 6-B as described in the chapter for ATC # 352.

Listed below are "Travel Air" model 10-D entries as gleaned from registration records: NC-676K ; Model 10-D (# 10-2001)

Wright R-760.
NC-368M ;   ,,   (# 10-2002)   ,,

;	,,	(# 10-2003)	,,
NC-694K ;	,,	(# 10-2004)	,,
NC-693K ;	,,	(# 10-2005)	,,
NC-150V ;	,,	(# 10-2006)	,,
NC-415N ;	,,	(# 10-2007)	,,
;	,,	(# 10-2008)	,,
NC-416N ;	,,	(# 10-2009)	,,
NC-471W ;	,,	(# 10-2010)	,,
NC-418N ;	,,	(# 10-2011)	,,

Registration numbers for serial # 10-2003 and # 10-2008 unknown.

Fig. 247. Performance and rugged character of Travel Air 10-D ideal for bush country.

Fig. 248. Lincoln Trainer model PT-K with Kinner K5 engine.

The Lincoln model PT-K was the next development in the PT (Page Trainer) series, a craft designed specifically for use by flying schools in their primary and secondary stages of flight training programs. Well balanced and well arranged aerodynamically the PT-K had long and easy moment arms to de-sensitize the pitching and bucking usually caused by an anxious or nervous hand on the "stick"; the amiable and gentle PT-K was a stable craft, easy to fly, and was quite forgiving in nature. In other words, the Lincoln PT-K was not too fussy and wouldn't get all upset due to a little pilot error. Performance with the added power and lowered gross weight was somewhat improved over the earlier PT and it was no exceptional trick for the PT-K to perform a well-rounded loop right from level flight; there was really no need to dive and gain excessive speed to bring it around. Though not fully acrobatic in the true sense, the lanky PT-K was sprightly and very versatile; it was able to perform easily most of the basic aerobatic maneuvers as taught in advanced pilot-training courses without much strain or a lot of coaxing.

The Lincoln model PT-K as shown here was an open cockpit biplane seating two in tandem and it was powered with the 5 cyl. Kinner K5 engine of 100 h.p. Like the earlier model PT, the PT-K was also kept bare of frills and fancy finery but not to the point of slab-sidedness; it presented a neat and trim appearance that would also be of some interest to the average pilot-owner as a sport-plane. The PT series, first brought out in an OX-5 powered version, achieved early success and were used as standard equipment on the flight-line by several large flying schools; the service record logged by these craft and the reputation they left behind was quite commendable. The type certificate number for the Lincoln model PT-K trainer, as powered with the 100 h.p. Kinner K5 engine, was issued 12-4-29 and some 18 or more examples of this model were manufactured by the Lincoln Aircraft Co., Inc. at Lincoln, Nebraska. Victor H. Roos was the president; Ray Page was V.P. in charge of sales and purchasing; and Ensil Chambers had replaced A. H. Saxon as the chief engineer. The term "Lincoln-Page" had since been dropped in favor of just plain "Lincoln" but flying-folk the country over had become familiar with "Lincoln-Page" over the years and the name was not too easily forgotten.

Listed below are specifications and performance data for the Lincoln trainer model PT-K as powered with the 100 h.p. Kinner K5 engine; length overall 25′7″; hite overall 9′0″; wing span upper 32′3″; wing span lower 31′9″; wing chord both 58″; wing area upper 154 sq.ft.; wing area lower 143 sq.ft.; total wing area 297 sq.ft.; airfoil Goettingen 436; wt. empty 1176 lbs.; useful load 591; crew wt. with 28.5 gal. fuel was 338 lbs. with 50 lb. baggage allowance (minus 20 lb. for each parachute carried); gross wt. 1767 lbs.; max. speed 104; cruising speed 85; landing speed 35; climb 800 ft. first min. at sea level; ceiling 13,500 ft.; gas cap. 28.5 gal.; oil cap. 4 gal.; cruising range at 7 gal. per hour was 330 miles; price at the factory field was $4175. later lowered to $3865. or $2235. less the engine and propeller.

The fuselage framework was built up of welded chrome-moly steel tubing in a modified Warren truss form, lightly faired to shape with wooden fairing strips and covered in fabric. The cockpits of 24x42 in. dimension were lightly upholstered and baggage allowance of 50 lbs. was carried in a compartment behind the rear seat; the bucket seats had wells for a parachute pack and the 2 parachutes at 20 lbs. each were part of the baggage allowance. The fuel tank was mounted high in the fuselage just ahead of the front cockpit, with a direct-reading fuel gauge projecting through the cowling. The wing framework was built up of solid spruce spar beams with spruce and basswood truss-type wing ribs; the completed framework was covered with fabric. There were 4 ailerons that were connected together in pairs by a streamlined push-pull strut. The landing gear of 76 inch tread was of the normal split-axle type using spools of rubber shock-cord (bungee) to absorb the bumps; wheels were 26x4 and no brakes were provided. The PT-K trainer was also eligible to operate on skis. The landing gear on all models of the PT series was later approved as a stiff-legged rigid gear, without rubber cord shock absorbers when using low pressure airwheels; the "airwheels" had enough "give" in the tire carcass due to the low air pressure used and therefore additional snubbing was not really necessary. The tail-skid was of the spring-leaf type with a removable hardened "shoe". The fabric covered tail-group was built up of welded chrome-moly steel tubing; the fin was ground adjustable and the horizontal stabilizer was adjustable in flight. Some of the model PT-K later had a light-framed canopy enclosure over the cockpits for bad weather protection. A wooden Supreme propeller was standard equipment on the PT-K but a metal propeller was optional, as were navigation lights, engine starter, tool kit, and Bloxham "safety-sticks".

Fig. 249. Rugged structure and compatible nature of PT-K ideal for pilot-training.

Normal color scheme for the PT-K was a medium blue fuselage with orange stripe and orange or orange-yellow wings. The next development in the PT trainer series was the Warner powered Lincoln PT-W as discussed in the chapter for ATC # 284 in this volume.

Listed below are Lincoln model PT-K entries as gleaned from registration records:

NC-35W;	Lincoln PT-K	(# 328)	Kinner K5.
;	„	(# 330)	„
NC-41N;	„	(# 601)	„
NC-275N;	„	(# 602)	„
NC-276N;	„	(# 603)	„
NC-277N;	„	(# 604)	„
NC-12204;	„	(# 605)	„
NC-534V;	„	(# 606)	„
NC-520V;	„	(# 607)	„
NC-521V;	„	(# 608)	„
NC-7W;	„	(# 609)	„
NC-8W;	„	(# 610)	„
NC-36W;	„	(# 611)	„
NC-37W;	„	(# 612)	„
NC-146W;	„	(# 613)	„
NC-407V;	„	(# 614)	„
NC-412V;	„	(# 615)	„
NC-408V;	„	(# 616)	„
;	„	(# 617)	„
NC-421V;	„	(# 954)	„

Serial # 328 and # 330 first as PT with OX-5 engine, also eligible as PT-K; registration number for serial # 330 and # 617 unknown; serial # 954 first as model PT-T, also eligible as PT-K.

Fig. 250. Travel Air model 4-P with 140 H.P. A.C.E. La. 1 engine; only one was built.

The model 4-P was the last and probably the rarest production version of the standard 3 place open cockpit "Travel Air" biplane, barring of course a few "specials" that were certificated from time to time on Group 2 approvals, and the 4-P was to mark the closing on the development of a series of airplanes that ran through at least 20 different models in the 5 years they were produced. Particularly adaptable to the installation of various engines, both water-cooled and air-cooled, in a power range that spread from 90 h.p. clear on up to 300 h.p., the "Travel Air" general-purpose biplane was a series that harbored several outstanding models that were all-time favorites in this period of aviation development; several of these models lasting in favor and popularity well on through the next decade. The "Model 1000" tried out its wings over the plains of Kansas in early 1925 as the curtain-raiser and the model 4-P had the distinction of bringing down the curtain as the last of a familiar and loveable breed of airplane. Just a year or so hence, an entirely new breed of "Travel Air" biplane was to grace the skies in numbers but they in no way detracted from the remembrance of the 2000, 3000, 4000, and the like.

The model 4-P (also as 4-PT) was a 3 place open cockpit biplane of the type that was currently coming off the Travel Air Co. line and it was powered with the newly introduced 7 cyl. A.C.E. model La. 1 engine of 140 h.p.; as shown, the engine was neatly enclosed in a speed-ring cowling that was a modified form of the N.A.C.A. low-drag engine fairing. Little is known about the characteristics or personality of the 4-P because the model certainly had no chance to acquire a country-wide reputation but we can easily assume that it must have shared all of the pleasant qualities of the "Travel Air" biplane. With fuel capacity held at 42 gal. maximum and the use of the lighter standard split-axle landing gear, it is apparent that it must have been calculated to keep the gross weight down on this version to offer a more sprightly performance without resorting to excessive horsepower. The 4-P was arranged more like a work-a-day airplane for the small operator and not particularly leveled at the sportsman-pilot; it is our guess that under more favorable conditions, this would have become a very popular model.

The A.C.E. model La. 1 was a development of the earlier Jacobs & Fisher engine (radial air-cooled) that was quite similar in appearance at least to the German "Siemans-Halske"; the A.C.E. La. 1 engine was developed further into the popular "Jacobs", a series of engines that had powered most of the popular makes of airplanes in this country for the next 10 years or so. The "Travel Air" model 4-P had much to offer in reliability, economy, and performance but it had been introduced to the public at a rather bad time, consequently, it was destined to remain a rare one-only type. A Group 2 approval numbered 2-160 (issued 12-4-29) was first awarded to the model 4-P for serial # 1332; a type certificate number for this same model and the same airplane was issued on 12-12-29. Only one example of this model was manufactured by the Travel Air Co. at Wichita, Kansas.

Listed below are specifications and performance data for the "Travel Air" model 4-P as powered with the 140 h.p. A.C.E. La. 1 engine; length overall 24'6"; hite overall 8'11"; wing span upper 33'0"; wing span lower 28'10"; wing chord upper 66"; wing chord lower 56"; wing area upper 171 sq.ft.; wing area lower 118 sq.ft.; total wing area 289 sq.ft.; airfoil "Travel Air"; wt. empty 1531 lbs.; useful load 857; payload with 42 gal. fuel was 392 lbs.; gross wt. 2388 lbs.; max. speed 115;

cruising speed 97; landing speed 45; climb 700 ft. first min. at sea level; ceiling 12,000 ft.; gas cap. 42 gal.; oil cap. 6 gal.; cruising range at 7.5 gal. per hour was 485 miles; baggage allowance was 50 lbs.; price at the factory field was $6240.

Construction details and general arrangement of the model 4-P was more or less similar to various other models in the "Travel Air" biplane series; we suggest reference to the chapters for ATC # 112 - 148 - 149 - 188 in U.S. CIVIL AIRCRAFT, Vol. 2 which discusses details and arrangement that will also apply to this model. A metal propeller, Bendix wheels and brakes, and speed-ring engine cowl were standard equipment. The next "Travel Air" development was the Curtiss-Wright 6-B cabin monoplane as discussed in the chapter for ATC # 352.

Listed below is the only known example of the "Travel Air" model 4-P:
NC-419N; Model 4-P (# 1332) A.C.E. La. 1-140.

Fig. 251. Rare model 4-P was the last in the series of Travel Air biplanes.

Fig. 252. Majestic Fokker F-32 with 4 "Hornet" engines; ship shown was last of 7 built.

The majestic "F-32" was to be "Tony" Fokker's largest and most elaborate creation; it was the "world's largest air transport" and was painstakingly designed to be the ultimate in air transportation at this time. With the spacious cabin divided into four separate compartments, the F-32 could seat 30 passengers and a crew of two as a day-coach transport in luxurious comfort; food was served in flight from the galley and serving pantry and the passengers also had complete lavatory facilities. As a night flying sleeper-plane it carried 16 passengers in pullman-type beds; there was a crew of 5 to see to their comforts and needs. It is evident the huge F-32 had been well thought out with a happy balance of utility, comfort, and performance but as we know of the sad story now, the airplane was far ahead of its need and was introduced at the wrong time. The prototype F-32, powered with 4 Pratt & Whitney "Wasp" engines, was built for Universal Air Lines who had 5 of these big craft on order and its introduction to the public was a heralded event in Sept. of 1929; in Nov. of 1929 on its maiden flight, it crashed among the houses when the port engines failed on take-off. Luckily no one was hurt and damage was not extensive but Universal was sure they didn't want the F-32 and backed out of their contract. Realizing the lack of reserve power,

the 450 h.p. "Wasp" engines were hastily replaced in the design with Pratt & Whitney "Hornet" B engines of 575 h.p. each; the extra 500 h.p. helped the F-32 a good deal and raised the performance to a phenomenal level for a craft of this size.

Western Air Express, who eventually had five of the F-32, launched passenger service with the huge craft from Los Angeles to San Francisco in April of 1930; there was an initial spurt of business because of the novelty but patronage of this service was certainly not as expected. "Pop" Hanshue of Western Air Express was heralding loud and often the fact that they were the proud possessors of these mammoth transports, service was extended to different parts of the west but they were soon sorely disappointed to find that passengers to fill all of those seats on each scheduled flight were sadly lacking; the extremely high overhead expenses on these flights was making it a notable but very unprofitable venture. Doggedly hanging on with trust and fierce determination, W.A.E. did the best they knew how to make it pay but it was a losing battle; the F-32 were then finally retired from active service as another interesting milestone in the progress of air transportation. Determined to prolong the breed as long as possible, a special F-32 was

Fig. 253. Introduction of prototype F-32 was occasion of great interest.

plushly outfitted as a "flying office" for L. P. Fisher, the president of the Cadillac Motor Car Co.; the inimitable "Tony" Fokker also launched several country-wide flights as promotion but this was all to no avail. The airplane was just too big and too expensive to be practical for these times.

Of the seven Fokker F-32 that were built, the first encountered near-disaster, the second after a number of months of unprofitable service, was sawn up and salvaged for useable parts, the third was also pulled from service and left standing around hopefully until it finally wound up as a "gas station" in Los Angeles. A Cleveland millionaire was negotiating to buy an F-32 to fulfill his dream and long-cherished plans of flying around the world in a westerly direction, in 10 days; records show no such attempt so we can assume the deal was dropped. One F-32 was tested by the Army Air Corps in 1931 as the YC-20 troop-transport but this was for test only and none were ordered. Disposition of the others left is unknown but the F-32 quickly faded from the scene; this was surely an untimely end for such a valiant effort and substantiates the feeling one gets that the fates were very often unkind to those who were bold enough to pioneer.

In manner of construction and general configuration the model F-32 (actually an F-12 in Fokker model designations but called the F-32 for its seating capacity) was a typical Fokker monoplane. As a high winged mono-

plane, the F-32 had four "Hornet" B engines of 575 h.p. each in nacelles on each side of the fuselage that were suspended from the huge cantilever wing; the engines were mounted in tandem, back to back. The front engines were tractor mounted and swung a 2-bladed metal propeller; the rear engines were mounted as "pushers" and swung a 3-bladed metal propeller to hold down the diameter for clearance. To promote more efficient cooling and additional speed, the engines were shrouded with low drag "speed-ring" cowlings; the huge wheels were covered with streamlined "pants" to improve the airflow and it is surprising that the F-32, as big as it was, could certainly keep up with the best of them. The type certificate number for the Fokker F-32, as powered with 4 "Hornet" B (R-1860) engines of 575 h.p. each, was issued 12-13-29 and some 7 examples of this model were manufactured by the Fokker Aircraft Corp. at Hasbrouck Hts. and Passaic, New Jersey. In May of 1930, the Fokker Aircraft Corp. which was already affiliated with the General Motors Corp., was taken over into the newly organized General Aviation Corp. and in Oct. of 1931 the name of the Fokker company was changed to the General Aviation Mfg. Corp. J. M. Schoonmaker was the president and general manager; Eddie V. Rickenbacker was director of promotion and sales; A. H. G. "Tony" Fokker was retained as the engineering director; and Herbert V. Thaden was the chief engineer. Robert Noorduyn, Fokker's right-hand man for many years, had already gone to work for Guiseppe

Bellanca.

Listed below are specifications and performance data for the "Fokker" model F-32 as powered with 4 "Hornet" B engines; length overall 69'10"; hite overall 16'6"; wing span 99'0"; max. wing chord 18'; total wing area 1350 sq.ft.; airfoil "Fokker"; wt. empty 14,910 lbs.; useful load 9340; payload with 400 gal. fuel was 6280 lbs.; payload with 700 gal. fuel was 4480 lbs.; gross wt. 24,250 lbs.; max. speed 146; cruising speed 123; landing speed 60; climb 1000 ft. first min. at sea level; climb in 10 min. was 7000 ft.; ceiling 13,500 ft.; gas cap. normal 400 gal.; gas cap. max. 700 gal.; oil cap. 40 gal.; cruising range at 112 gal. per hour was 430-740 miles; price at the factory was $110,000. The prototype F-32 weighed in at 13,800 lbs. empty, had a useful load of 8100 lbs., and gross wt. was 22,500 lbs.; the latest versions of the F-32 weighed in at 15,080 lbs. empty, had a useful load of 9170 lbs., payload with 480 gal. fuel was 5630 lbs.; and gross wt. was 24,250 lbs.

The huge fuselage framework was built up of welded chrome-moly & mild carbon steel tubing; the framework was faired to shape with wooden bulkhead formers, wooden fairing strips and then fabric covered. The floors in the main cabin and pilots cockpit area were of corrugated dural and plywood sheet that was supported in the framework by riveted aluminum girders; bracing in the cabin area was so arranged to leave an unobstructed center aisle as passage-way to all compartments. The enclosed pilots cockpit up front had dual swing-over control wheel and was reached either through a door leading from the main cabin or through a door in the floor of the nose section; the baggage compartment and radio room were below the

floor of the cockpit. Earlier versions of the F-32 had a 120 cu. ft. compartment with allowance for 1350 lbs. of baggage, while later versions were cut down to 85 cu. ft. and allowance for 660 lbs. of baggage; first allowance was 45 lbs. of baggage per passenger and later was cut to 22 lbs. per passenger. The main cabin area was divided into 4 compartments with a small galley and serving pantry in the center; behind the last compartment was the cabin lobby with entry through two large doors and to the rear of this area were the two lavatories. Passenger seats were arranged in pairs on each side of the central passage-way; the cabin was heated from the engine-exhaust system, there were adjustable vents for ventilation and cabin lights were provided for passenger convenience. Each cabin compartment had 4 large windows of shatter-proof glass on each side, the inner walls were lined with balsa-wood for insulation and sound-proofing, and the seats were exceptionally large and comfortable. The pilots cabin was of 50x60x60 inch dimension and the main cabin was of 100x350x70 inch dimension of clear area. The huge cantilever wing framework which tapered towards the tips in plan-form and in thickness, was built up of laminated spruce spar beams of box-type construction which were 44 inches at their maximum depth; wing ribs were of spruce flanges and plywood webs, plywood stringers provided additional bracing, and the completed framework was covered with plywood veneer. The wing was bolted directly on top of the fuselage and the main fuel tanks were mounted in the center portion, just above and inboard from the engine nacelles; the oil tanks, one for each engine, were mounted in the center of each nacelle. The engine mountings were welded chrome-moly steel tube structures suspended from the wing spars by braces of

Fig. 254. First Fokker F-32 shown during final assembly at factory.

Fig. 255. Impressive view of F-32; rear engines swung "pusher" propellers.

streamlined steel tubing and covered with detachable cowling panels; the treadle type landing gear of 240 inch tread was fastened to each nacelle structure and used Gruss oleo-spring shock absorbing struts. The main wheels were 58x14 with Bendix brakes; the tail-wheel was steerable to 30 deg. either side and could be unlocked to swivel freely in ground maneuvering. The fabric covered tail-group of "balanced" rudders and elevators was built up of welded chrome-moly steel tubing; the proto of the F-32 had twin rudders but later examples had triple rudders for better directional control. The engines were provided with electric inertia-type engine starters and engine-driven generators were mounted on each front engine; pressure type fire extinguishers were mounted in the engine compartments and operated from the cockpit. The Fokker F-32 was the biggest and most elaborate commercial airplane to be built by A. H. G. "Tony" Fokker in this country, and it was also the last.

Listed below are Fokker model F-32 entries as gleaned from registration records:

X-124M;	Fokker F-32	(# 1201)	4 Wasp.
NC-130M;	„	(# 1202)	4 Hornet.
NC-333N;	„	(# 1203)	„
NC-334N;	„	(# 1204)	„
NC-335N;	„	(# 1205)	„
NC-336N;	„	(# 1206)	„
NC-342N;	„	(# 1207)	„

Fig. 256. First of F-32 super-transport for Western Air Express; greeted here at St. Louis while on flight to California.

Fig. 257. American Eagle "Phaeton" model R-540 with 165 H.P. Wright J6 engine.

Ever since its introduction in early 1926, the popular "American Eagle" biplane as first designed by R. T. "Bob" McCrum, had remained more or less the same in general configuration except for some refinements to the landing gear, tail-group and aileron arrangement. These changes were just logical developments to a basic design that was flexible and versatile, and did lend itself to modification with any number of engine installations for a varied number of uses. The standard engine installation had long been the Curtiss OX-5 and in 1929 it was the 5 cyl. Kinner; in between these two there were models offered with at least 5 different engines, all on the same basic airframe. For 1930, an entirely new concept had been planned by Ed Porterfield and Jack Foster, whereby each new model was to be designed and arranged more specifically for a certain purpose, with an engine and configuration carefully selected to suit. The "Phaeton", or model 39, was the first in a completely new line of airplanes to be introduced by "American Eagle" for 1930.

Under the supervision of Larry Ruch and Jack Foster, the first American Eagle "Phae-

ton" was already in a testing program during June-July of 1929; results were very satisfactory and the first in a new series was soon scheduled for quantity production. As shown here, the new "Phaeton" was a 3 place open cockpit biplane of the general-purpose type and though it did not resemble the earlier "American Eagle", its characteristics and general behavior were more or less the same in many respects; overall there was a general improvement and better performance for the amount of power expended. As the model R-540 the "Phaeton" was powered with the 5 cyl. Wright J6 (R-540) engine of 165 h.p. and was basically planned for use by the flying services for flight instruction and air-taxi work. Some noticeable changes in the new "Eagle" were slight overhang in the upper wing, a center-section panel supported by N-type struts, redesigned tail-group and a long-leg landing gear of wide tread; the beefed-up structure of the "Phaeton" was no doubt planned early for much higher horse-power. The "Phaeton" was a fairly good airplane but it was surely not what the flying public was buying in 1930, so it remained quite scarce in number. The first "Phaeton",

Fig. 258. The "Phaeton" featured new lines for 1930.

as the model 39, was certificated on a Group 2 approval numbered 2-123 (issued 9-7-29); the type certificate number for the "Phaeton" as powered with the 165 h.p. Wright J6 engine was issued 12-13-29 and this superseded the Group 2 approval. Some 8 or more examples of the model R-540 (also as model 251) were manufactured by the American Eagle Aircraft Corp. at Fairfax Airport in Kansas City, Kansas. Ed. E. Porterfield, Jr. was president; D. H. Hollowell was sales manager; Jack E. Foster was chief engineer; and Larry D. Ruch was chief pilot in charge of test and development.

Listed below are specifications and performance data for the American Eagle "Phaeton" model R-540 as powered with the 165

h.p. Wright J6 engine; length overall 24'5"; hite overall 8'5"; wing span upper 31'0"; wing span lower 28'10"; wing chord both 63"; wing area upper 159 sq.ft.; wing area lower 134 sq.ft.; total wing area 293 sq.ft.; airfoil USA-35B; wt. empty 1759 lbs.; useful load 801; payload with 42 gal. fuel was 344 lbs.; gross wt. 2560 lbs.; max. speed 120; cruising speed 100; landing speed 42; climb 816 ft. first min. at sea level; ceiling 14,000 ft.; gas cap. normal 42 gal.; gas cap. max. 60 gal.; oil cap. 5 gal.; cruising range at 9 gal. per hour was 390-560 miles; price at the factory field was $6795., soon lowered to $5995.

The fuselage framework was built up of welded chrome-moly steel tubing in a rigid truss form, faired to shape with wooden fairing strips and fabric covered. The cockpits were deep and well protected with a large door on the left side for entry to the front cockpit. The wing framework was built up of solid spruce spar beams that were routed out for lightness with spruce and plywood truss-type wing ribs; the leading edges were covered with plywood and the completed framework was covered with fabric. The main fuel tank was mounted high in the fuselage just ahead of the front cockpit; extra fuel was carried in a gravity-feed tank mounted in the center-section panel of the upper wing. Extra fuel

Fig. 259. Clipped-wing "Phaeton" used for racing.

Fig. 260. Larry Ruch flying "Phaeton" during early tests. Prototype awarded group 2 approval.

brought max. fuel load to 60 gal. and this cut 108 lbs. from payload allowance. The center-section and wing bracing struts were N-type structures of chrome-moly steel tubing in streamlined section; the 4 ailerons were connected together in pairs by a streamlined push-pull strut. The split-axle landing gear of 86 inch tread was of the long-leg type and used oil-draulic shock absorbing struts; wheels were 28x4 or 30x5 and brakes were standard equipment. A metal propeller and wiring for navigation lights was also standard equipment; an inertia-type engine starter was optional equipment. The next development in the "Phaeton" biplane was the model R-760 as described in the chapter for ATC # 283

in this volume.

Listed below are "Phaeton" model R-540 entries as gleaned from registration records:

C-872E; Phaeton R-540 (# 550)

Wright R-540.

NC-516H;	,,	(# 551)	,,
NC-543H;	,,	(# 601)	,,
NC-560H;	,,	(# 602)	,,
NC-561H;	,,	(# 603)	,,
NC-562H;	,,	(# 604)	,,
NC-572H;	,,	(# 605)	,,
NC-577H;	,,	(# 606)	,,

Serial # 550 and # 551 also had Wright J5 engine, eligible for ATC # 283; serial # 604 (NR-562H) also as clipped-wing racer.

Fig. 261. "Phaeton" harbored many changes over earlier "Eagle."

A.T.C. #283
(12-16-29)
AMERICAN EAGLE "PHAETON", J5

Fig. 262. American Eagle "Phaeton" model J-5 with 220 H.P. Wright J5 engine.

Somewhat contrary to previously announced plans that stated each new "American Eagle" was to be designed specifically for a certain purpose, with engine and configuration to match, the "Phaeton" biplane was also offered with the "Hisso" (Hispano-Suiza) engines, the Wright "Whirlwind" J5 and the 7 cyl. Wright J6 engine; it would seem that 3 place open cockpit biplanes of this type were just naturals for modification with various engine installations and the temptation to do this was just too hard to resist. As a matter of fact, the "Phaeton" biplane was quite an airplane with any of the powerplants mentioned above; 180 h.p. to 225 h.p. was surplus power that translated into plenty of good performance. 'Twas often said that president Ed Porterfield, as a pilot, made rigid demands of an airplane and he naturally wanted these sought-after qualities incorporated into every "American Eagle" that left the plant.

The American Eagle "Phaeton" model J5 with the 9 cyl. Wright "Whirlwind" J5 engine of 220 h.p. was also a 3 place open cockpit biplane of the general-purpose type but this

model was especially leveled at the sportsman-pilot who would enjoy a capable ship with all the extras in performance. The "Phaeton" as the model R-760 was also offered with the Wright "Whirlwind Seven" engine of 225 h.p.; both models were comparable in characteristics and performance so selection of either would be but a matter of engine preference. During the 1930 National Air Races held in Chicago, Miss Jean LaRene flew a "Phaeton" model R-760 to 3rd place in the class A Pacific Derby for women. Either the "J5 Phaeton" or the "Model R-760" were airplanes well suited to the requirements of the sportsman-pilot but these sportsman-pilots were at a minimum during this particular period so only one of each example were built. The model R-760 was certificated on a Group 2 approval numbered 2-121 (issued 9-6-29) for serial # 550; on the next day (apparently after a quick engine change) this same airplane was certificated as the model R-540 on a Group 2 approval numbered 2-123 with the 5 cyl. Wright J6 engine. The type certificate number for the "Phaeton J5" was issued 12-16-29 and later amended to include the model R-760; this superseded Group 2 approval numbered 2-121. One each of the "Phaeton J5" and "Phaeton R-760" were manufactured by the American Eagle Aircraft Corp. at Fairfax Field in Kansas City, Kansas. American Eagle's new plant on Fairfax Field now displaced 60,000 feet of floor area and was designed for large scale production.

Listed below are specifications and performance data for the American Eagle "J5 Phaeton" as powered with the 220 h.p. Wright J5 engine; length overall 23'6"; hite overall 8'8"; wing span upper 31'0"; wing span lower 28'10"; wing chord upper & lower 63"; wing area upper 159 sq.ft.; wing area lower 134 sq.ft.; total wing area 293 sq.ft.; airfoil USA-35B; wt. empty 1896 lbs.; useful load 907; payload with 42 gal. fuel was 450 lbs.; payload with 60 gal. fuel was 342 lbs.; gross wt. 2803 lbs.; max. speed 130; cruising speed 110; landing speed 47; climb 1000 ft. first min. at sea level; ceiling 15,000 ft.; gas cap. normal 42 gal.; gas cap. max. 60 gal.; oil cap. 5 gal.; cruising range at 12 gal. per hour was 385—530 miles; price at the factory field was $7795. Specifications and performance figures for the "Model R-760" were comparable except for a slight difference in weights; gross wt. was 2789 lbs. and price was slightly higher.

The construction details and general arrangement was more or less the same for "Phaetons" of all models; for comparison, we suggest reference to the chapter for ATC # 282 in this volume. The "J5 Phaeton" had 30x5 wheels with Bendix brakes, a metal propeller, and bayonet-type engine exhaust stacks; an inertia-type engine starter and navigation lights were optional equipment. The model R-760 was equipped in similar fashion except for exhaust-collector ring as used with all Wright J6 series engines. All "American Eagle" models could now be purchased on the time payment plan with a down payment and monthly payments usually tailored to suit the individual requirements. The next development in the "American Eagle" biplane was the "Model 201" as described in the chapter for ATC # 293 in this volume.

Listed below are the only known entries of the "Phaeton" as eligible for this certificate: X-872E; Phaeton J5 (# 550) Wright J5. X-516H;   ,,   (# 551)   ,,

Serial # 550 and # 551 both also had "Whirlwind Seven" as model R-760; serial # 550 also converted to model R-540 with "Whirlwind Five".

Fig. 263. The "Phaeton" was also powered with "Hisso" (Hispano-Suiza) engines.

Fig. 264. Lincoln Trainer model PT-W with 110 H.P. Warner engine; hailed as the best of the PT series.

Like the Lincoln PT-K described here previously, the model PT-W was a progressive development in the series of "Page Trainers", an open cockpit biplane that was typical in most all respects except for its powerplant installation and whatever modifications were necessary to accommodate the 7 cyl. Warner "Scarab" engine. The basic characteristics inherent in the Lincoln PT design were quite ideal for flight training purposes and the model PT-W was hailed as probably the best craft in the series. It was described as a stable and very docile craft that allowed an unusual amount of error in judgement, timidity at the controls, or misuse of piloting technique before it became somewhat disturbed and near-to uncontrollable; these were certainly appreciated qualities well suited to an airplane used extensively for pilot training in all of its phases. Very responsive to "stall" and "spin" recovery, it was many times said that either the PT-K or the PT-W, when fitted with soft cushion low-pressure "airwheels", could be actually mushed in from 50 feet in

a stalled condition and not even damage the landing gear; no doubt a good pilot could do it but whether a student pilot could perform this feat is open to some question.

Quite a rare version in the Lincoln PT series, the model PT-W was a lean and tidy looking open cockpit biplane that seated two in tandem and was powered with the 110 h.p. Warner engine. Though primarily designed as a pilot-training airplane the PT-W did have certain characteristics of behavior and performance that were well suited for sport-flying as practiced by the average pilot-owner; its fairly high price tag for this particular lean period in aviation no doubt kept it from being produced in greater number. The model PT-W was also eligible to operate on skis for winter flying and later modifications to this design allowed the PT-W to operate with a stiff-legged rigid landing gear when using low pressure "airwheels". The type certificate number for the model PT-W as powered with the 110 h.p. Warner "Scarab" engine was issued 12-31-29 as the last number issued for the year 1929 and some 4 or more examples of this model were manufactured by the Lincoln Aircraft Co., Inc. at Lincoln, Nebraska. In the chapter for ATC # 28 of U.S. CIVIL AIRCRAFT, Vol. 1, we mentioned the fact that the Lincoln-Page model LP-3 was almost an exact copy of the "OX-5 Swallow" and could not find any reasonable explanation for the obvious imitation. It has since been learned that Lincoln-Page did have a valid reason for the imitation and this was not generally known. Old files bear a statement that tells of "Jake" Moellendick selling the rights to manufacture the "Swallow" biplane to the Lincoln-Page group; this deal was to have been transacted less than a month previous to the time the Swallow Airplane Co. went into receivership and was taken over by another group. The new group apparently knew nothing of this earlier transaction and resented openly the obvious imitation by Lincoln-Page intensely. Victor H. Roos who had formerly been with the Swallow Airplane Co., was now the president of Lincoln Aircraft Co.; Ray Page was V.P. in charge of sales and purchasing; and Ensil Chambers was the chief engineer.

Listed below are specifications and performance data for the Lincoln model PT-W as powered with the 110 h.p. Warner "Scarab" engine; length overall 25'3"; hite overall 9'0"; wing span upper 32'3"; wing span lower 31'9"; wing chord both 58"; wing area upper 154 sq.ft.; wing area lower 143 sq.ft.; total wing area 297 sq.ft.; airfoil Goettingen 436; wt. empty 1203 lbs.; useful load 591; crew wt. with 28.5 gal. fuel was 338 lbs.; baggage allowance 50 lbs.; gross wt. 1794 lbs.; max. speed 108; cruising speed 87; landing speed 36; climb 870 ft. first min. at sea level; ceiling 14,000 ft.; gas cap. 28.5 gal.; oil cap. 4 gal.; cruising range at 7 gal. per hour was 330 miles; price at the factory was first quoted at $4625. but soon lowered to $4315.; also available for $2235. less engine and propeller.

Construction details and general arrangement for the Lincoln model PT-W were typical of the model PT-K as described in the chapter for ATC # 279 in this volume. The cockpits of 24x42 inch dimension had bucket-type seats with wells for a parachute pack; windshields were rather large to offer good weather protection. Engine and flight instruments were mounted in both cockpits, while the ignition switch, fuel shut-offs, starting choke, cowling shutters, and horizontal stabilizer adjustment were controllable from either cockpit. Wheels were 26x4 and no brakes were provided; low pressure airwheels were available for $133.35 extra. The model PT-W was also eligible to operate on skis and later modifications to this design allowed the use of a stiff-legged rigid landing gear when using the soft cushion "airwheels"; this eliminated the rubber cord shock absorbers. A wooden Supreme propeller was standard equipment but a metal propeller was available; other equipment included tool kit, Bloxham safety sticks, and a fuel gauge. The next development in the Lincoln PT series was the Brownback "Tiger" powered model PT-T as discussed in the chapter for ATC # 344.

Listed below are Lincoln model PT-W entries as gleaned from registration records: NC-561M; Model PT-W (# 801) Warner 110.

NC-426V;	,,	(# 802)	,,
;	,,	(# 803)	,,
;	,,	(# 804)	,,
-11086;	,,	(# 851)	,,

Registration numbers for serial # 803 and # 804 unknown; serial # 851 was model PT-M with Martin 333 engine, also eligible as PT-W; all serial numbers also eligible with 125 h.p. Warner engine.

Fig. 265. "Courier" model PB-1 with Kinner K5 engine; craft was excellent for all-purpose work.

The "Courier" model PB-1 was a 3 place high winged cabin monoplane of robust proportion, simple straight-cut lines, and of a very plain every-day configuration; it was an airplane that didn't seem to impress one at first as outstanding for any particular purpose but it did look like it would do an honest job for anyone without coaxing or coddling. Planned primarily for general-purpose work, the "Courier" monoplane was supposed to handle all chores usually performed by the 3 place open cockpit biplane, plus the bonus of cabin comfort and somewhat higher performance; there's no doubt it did fulfill the promise as planned but the modest PB-1 sort of got lost in the shuffle and remained virtually unknown and a very scarce and rare type.

During a casual meeting with a fellow-worker some years ago, we lightly began discussing airplanes because this is what brought us together; touching briefly on several of the better-known types we were both familiar with through the years, my new-found acquaintance stated that he once owned a fine little airplane that I had probably never heard of before. Bent on further details, we inquired anxiously as to what it could be and he stated almost proudly that it was a "Courier" monoplane with a 5 cyl. Kinner K5 engine. We assured him that we had heard of it but had never actually seen one, and upon that note

we talked further; for a good half hour he had naught but praise for its character and its nature so from this we had to assume that the "Courier" monoplane was an airplane that was easy to get along with and would perform its duties, whether difficult or trite, with satisfaction and the minimum of fuss. This former "Courier" pilot-owner assumed his ship was the first and only example but actually there were at least three.

The Courier Monoplane Co. with a cubby-hole office in Wilmington, Calif., hoping to attract some financing, had an exhibit at the 1928 National Air Races held in Los Angeles; blueprints and descriptive brochures were shown for the proposed manufacture of various aircraft. With financing promised and secured the Courier Monoplane Co. was formally formed at Los Angeles in Dec. of 1928; tooling was hastily set up and work was begun on a prototype version. The "Courier" monoplane was designed by the well-known Wm. J. Waterhouse, a Los Angeles consulting engineer who was instrumental in the design and development of several fine airplanes built in the So. Calif. area. With the prototype airplane built and successfully tested in early 1929, operations had been moved over to Long Beach, Calif. in a plant-site formerly used by the International Aircraft Co.; space here was ample enough to accommodate the expansion that was eagerly planned for "Courier" monoplanes production. To promote local interest in their new craft, the "Courier" PB-1 was proudly exhibited at the State Fair in Sacramento, Calif. as the latest thing in vehicles for the progressive farmer. Primed and ready for a flood of orders, unfortunately there was only a trickle to be had so company engineers designed the "Courier" MT-1 trainer which was of similar character but done up as a "parasol" monoplane with two open cockpits. Neither the model PB-1 nor the MT-1 actually had a chance to prove their worth because of the economic uncertainty in this country at the time and before long the Courier Monoplane Co. slowly faded from the scene.

The "Courier" model PB-1 was a high winged cabin monoplane seating three with plenty of room and comfort; the pilot sat up front in a bucket-type seat where he had ample vision and the two passengers sat on a bench-type seat in back. Easy entry and exit was gained by two large doors on the right side and a baggage compartment was provided behind the rear seat. With 32 gal. of fuel aboard the cruising range was extended to nearly 450 miles but maximum fuel load allowed no baggage when carrying three. First approved on a Group 2 certificate numbered 2-131 (issued 9-26-29) the "Courier" model PB-1 as powered with the 100 h.p. Kinner K5 engine received its type certificate number on 1-3-30; some 2 or more examples of this model were manufactured by the Courier Monoplane Co. at Long Beach, Calif. L. J. Bailey was

president; M. J. Fix was V.P. and general manager; E. S. Plunkett was treasurer and sales manager; and Wm. J. Waterhouse was the chief engineer.

Listed below are specifications and performance data for the "Courier" model PB-1 as powered with the 100 h.p. Kinner K5 engine; length overall 24'6"; hite overall 7'8"; wing span 37'0"; wing chord 72"; total wing area 200 sq.ft.; airfoil Goetingen 398; wt. empty 1387 lbs.; useful load 713; payload with 32 gal. fuel was 320 lbs.; gross wt. 2100 lbs.; max. speed 105; cruising speed 88; landing speed 40; climb 650 ft. first min. at sea level; ceiling 12,000 ft.; gas cap. 32 gal.; oil cap. 4 gal.; cruising range at 6.5 gal. per hour was 440 miles; price at the factory was set at $5750. but soon lowered to $5500.

The fuselage framework was built up of welded chrome-moly steel tubing in a Warren truss form with an arch construction used in the cabin area to eliminate obstructions from diagonal bracing; the framework was faired to shape with spruce fairing strips and then covered in fabric. With an abundance of window area, including side-panels for the pilot, visibility was more than ample for an arrangement of this type; window panels could be slid open for ventilation and an exhaust-pipe muff provided cabin heat. The semi-cantilever wing framework, in two halves, was built up of box-type spar beams of spruce flanges and filler blocks with 3 plywood webs; the wing ribs were spruce and plywood trusses and the completed framework was covered with fabric. The trailing edges

Fig. 266. "Courier" Trainer model MT-1 with Kinner K5 engine; rugged simplicity was its forte.

and wing tips were formed of steel tubing, the leading edges were covered with plywood to preserve the airfoil form, and the nose of the balanced-hinge ailerons covered with aluminum sheet. The wing halves were bolted to the top longerons of the fuselage and braced by parallel struts; the gravity-feed fuel tank of 32 gal. capacity formed the top of the cabin area. The landing gear of 106 in. tread was of the out-rigger type and was built into a strong truss with the wing bracing struts; shock absorbers were Gruss oleo-spring struts, wheels were 26x4 and Kelsey-Hayes brakes were standard equipment. The fabric covered tail-group was built up of welded chrome-moly steel tubing; the fin was ground adjustable and the horizontal stabilizer was adjustable in flight. Contrary to average practice, the horizontal stabilizer was hinged on its leading edge and a worm-gear raised or lowered the aft portion for trimming with variable loads. The noisy Kinner engine was muffled by a large volume collector-ring and this kept cabin noise at a fairly low level; a tail-pipe deflected exhaust gases well below the cabin level. A wooden propeller, wheel brakes, navigation lights, and cabin lights were standard equipment; a metal propeller, dual controls, and engine starter were available as optional equipment.

Listed below are "Courier" model PB-1 entries as gleaned from registration records:

X-280E;	Courier PB-1	(# 100)	Kinner K5.
NC-181N;	,,	(# 101)	,,
NC-168W;	,,	(# 102)	,,

Serial # 100 was prototype airplane approved on Group 2 approval numbered 2-131. The 5 cyl. Kinner K5 engine was riding the crest of a wave of popularity at this time; 49 manufacturers of aircraft were using the K5 in 1929. Of course, all of these were not production models, several of these were experimentals and some prototypes that were built in only one or two examples.

A.T.C. #286
(1-2-30)
SPARTAN, C3-225

Fig. 267. Spartan biplane model C3-225 with 225 H.P. Wright J6 engine.

The staid and calm "Spartan" biplane was a well-behaved airplane that didn't seem to provoke too much national comment one way or another and just plied along in its quiet way, unobtrusive and without any fanfare in the multitude of chores it was usually asked to do. Like any other good general-purpose biplane it was used in an endless variety of tasks and it did them all very well; most comment about this craft seems to narrow down to the fact that it was a "good airplane" and we would be hard put to analyze the reason for this general under-statement. Previously powered with engines of 120-165 h.p. the C-3 series by Spartan delivered very good performance and the mild-mannered personality of these craft soon inspired a trust that nearly amounted to complacency. Continuously upgraded with latest improvements and with adherence to a rigid policy of good craftsmanship, the "Spartan" biplane was now offered in a version that was expressly leveled at the pilot-salesman, the pilot-executive, or the sportsman-pilot who was in search of somewhat better performance. Though still quite typical to the earlier C3-165 in many ways, the new model C3-225 was now powered with the 7 cyl. Wright J6 engine of 225 h.p.; this added power showed up as an increase in general performance with the ability to handle a fair amount of payload under just about any conditions. Along with the increase in power came a slightly different personality; though still no "tiger" by any means, the broad-chested C3-225 displayed plenty of verve and muscle to make it a good match for most any craft of this particular type. Most examples of the "Spartan" C3-225 were kept busy as air-taxis in the oil-fields of the middle west and one was used for a time in weather research; several of the C3-225 were still in active service up to 10 years later and at least one has been rebuilt to this configuration to fly again in recent years.

The buxom "Spartan" model C3-225 was an open cockpit biplane of the general purpose type with seating for three; arranged to offer the greatest amount of utility possible, the cockpits were handy, roomy, and well protected. For efficient and reliable service in the field, the C3-225 was of stout framework throughout to lessen maintenance, with helpful features for ease of service. In order to

Fig. 268. Handsome C3-225 was the finest example of the Spartan biplane.

offer that extra bit of performance and a corresponding increase in utility that was becoming essential in this type of airplane, the "Spartan" was now powered with the 7 cyl. Wright J6 (R-760) engine of 225 h.p.; with this amount of power the new C3-225 was comparable to other aircraft of this type and still more or less retained the compatible nature so typical of this series. Except for the one and only model C3-166, which also came out about this time, the model C3-225 was destined to ring down the curtain as the last of the "Spartan" biplanes; the usefulness of this type of craft, especially in this higher price range, was rapidly being narrowed and the flying public was looking to the enclosed monoplane more and more. The type certificate number for the "Spartan" model C3-225, as powered with the 225 h.p. Wright "Whirlwind Seven" engine, was issued 1-2-30 and some 14 examples of this model were manu-

factured by the Spartan Aircraft Co. at Tulsa, Oklahoma.

Listed below are specifications and performance data for the "Spartan" model C3-225 as powered with the 225 h.p. Wright R-760 engine; length overall 23'3"; hite overall 8'10"; wing span upper & lower 32'0"; wing chord both 60"; wing area upper 151 sq.ft.; wing area lower 140 sq.ft.; total wing area 291 sq.ft.; airfoil Clark Y; wt. empty 1741 lbs.; useful load 959; payload with 60 gal. fuel was 384 lbs.; gross wt. 2700 lbs.; max. speed 130; cruising speed 110; landing speed 55; climb 1160 ft. first min. at sea level; ceiling 15,000 ft.; gas cap. 60 gal.; oil cap. 6.5 gal.; cruising range at 13.5 gal. per hour was 460 miles; price at the factory field was $7750.

The fuselage framework was built up of welded chrome-moly steel tubing, heavily faired to shape with wooden formers and fairing strips, then fabric covered. The cockpits were deep, roomy, and well upholstered; a baggage compartment of 6.8 cu. ft. capacity with allowance for 40 lbs. was in the turtleback section of the fuselage just behind the rear cockpit. The wing framework was built up of solid spruce spars that were routed to an I-beam section with spruce and plywood truss-type wing ribs; the leading edges were covered with dural sheet and the completed framework was covered in fabric. There were four

Fig. 269. Spartan C3-225 used in high altitude weather research.

ailerons that were connected together in pairs by a push-pull strut; interplane struts were of streamlined chrome-moly steel tubing in a heavy gauge and interplane bracing was of heavy streamlined steel wire. The lower wing was faired neatly into the fuselage and all exposed wing fittings were covered with streamlined caps. The main fuel tank of 45 gal. capacity was mounted in the center-section panel of the upper wing; an extra fuel tank of 15 gal. capacity was mounted high in the fuselage just ahead of the front cockpit. The split-axle landing gear of 83 inch tread was of two long telescopic legs with oleo-spring shock absorbing struts; wheels were 30x5 and Bendix brakes were standard equipment. The fabric covered tail-group, with an increase in fin and rudder area, was built up of welded chrome-moly steel tubing; the fin was ground adjustable and the horizontal stabilizer was adjustable in flight. A liberal number of zippered inspection panels were provided for inspection of the control mechanisms and other important areas in the airplane. A Hamilton metal propeller, booster magneto, battery, navigation lights, and wheel brakes

were standard equipment. Dual controls, tail wheel, speed-ring engine cowling, windshield for pilot of shatter-proof glass panels, and hand-crank inertia-type engine starter were available as optional equipment. The next and the final development in the "Spartan" biplane was the model C3-166 as described in the chapter for ATC # 290 in this volume.

Listed below are "Spartan" model C3-225 entries as gleaned from registration records:

X-700N;	Spartan C3-225	(# A-1)	
			Wright R-760.
NC-708N;	„	(# A-2)	„
NC-709N;	„	(# A-3)	„
NC-710N;	„	(# A-4)	„
NC-711N;	„	(# A-5)	„
NC-712N;	„	(# A-6)	„
NC-713N;	„	(# A-7)	„
NC-714N;	„	(# A-8)	„
NC-715N;	„	(# A-9)	„
NC-716N;	„	(# A-10)	„
NC-717N;	„	(# A-11)	„
NC-718N;	„	(# A-12)	„
NC-719N:	„	(# A-13)	„
NC-720N;	„	(# A-14)	„

Fig. 270. Dependable utility of C3-225 popular among oil companies for liaison duty in oil-fields.

Fig. 271. Savoia-Marchetti model S-56 with Kinner K5 engine.

In nearly every city having the advantage of nearby water, were wealthy owners of motor-cars and speed-boats who were often chafing at the bit, so to speak, about the road congestion and bother of surface travel to popular resorts and playgrounds; these were the people singled out by the American Aeronautical Corp. as likely prospects for their new "baby amphibian". These were the people who would be ready and willing to take to the air to travel swiftly and unhampered in a small easy-to-handle amphibious aircraft especially designed for private flying. The Savoia-Marchetti model S-56 being a small craft of simple and rugged construction, was ideal and well suited for this type of all-purpose flying and would cost hardly no more to operate and maintain than the average speed-boat.

The wooden-hulled S-56 was a biplane of the flying boat type with a simple hand-operated retracting landing gear; seating for 3 in two open cockpits might have lacked the plush appointments of formal comfort but it did make up for it in fun and utility. Powered with a 5 cyl. Kinner K5 engine of 100 h.p., mounted under the top wing in a "tractor" fashion, the likeable S-56 had average performance for a craft of this type with maintenance and operating costs held to a sensible minimum. The pert model S-56 was an Italian design that was built under license from the Savoia Co. in Milan but several modifications were incorporated to suit American manufacturing and operating conditions. A small claim to fame was achieved by the diminutive S-56 when it became the first airplane to be actively used by a police force to help enforce flying regulations over heavily populated areas; the Police Dept of New York City led the way by establishing a flying division and had several policemen trained for flying duties on an "aerial beat".

The American Aeronautical Corp. was

Fig. 272. S-56 "baby" amphibian ideal for fun or sport.

formed late in 1928 by Enea Bossi to manufacture the Savoia-Marchetti models S-55, S-56 and S-62 in this country; designs born of proven experience since 1916. The unusual S-55 was a large twin hulled monoplane transport with 2 tandem-mounted engines that had already made a 60,000 mile tour of six continents in 1927 and later a number of other notable flights. The model S-62 was a medium capacity transport and mail-carrier; the baby S-56 is described above. A prototype version

of the S-56 was issued a Group 2 approval numbered 2-96 (issued 8-2-29) as a 2 place craft with Kinner K5 engine; two other test versions were issued a Group 2 approval numbered 2-95 also as 2 pl. craft with the Kinner K5 engine. Serial numbers 4 and upwards were issued an approved type certificate number on 1-4-30. In Feb. of 1930 it was reported that 14 of the model S-56 had already been built and one S-55 twin-hulled flying boat was nearing completion; altogether some

Fig. 273 Italian version of S-56 was powered with "Walter" or "Siemens" engines.

*Fig. 274. Unusual model S-55 was
twin-hulled transport.*

foil "Marchetti"; wt. empty 1451 lbs.; useful load 699; payload with 28 gal. fuel was 328 lbs., allowing 27 lbs. for baggage; gross wt. 2150 lbs.; max. speed 86; cruising speed 75; landing speed 40; climb 500 ft. first min. at sea level; ceiling 7000 ft.; gas cap. 28 gal.; oil cap. 3 gal.; cruising range at 7 gal. per hour was 290 miles; price at the factory ramp was $7300., later raised to $7375. with engine starter and navigation lights.

30 or more examples of the S-56 version were built with the Kinner K5 engine. To stimulate international interest and drum up export sales, an S-56 was sent to tour likely ports in So. America. The earliest prototype version of the S-56 was powered with a Siemens-Halske engine and a Walter engined version was reported also. The American Aeronautical Corp. with executive offices in New York City, had a factory and flight ramp on 12 acres of shore-front property in Port Washington, Long Island, N. Y. Ugo V. D'Annunzio was the president; P. G. Zimmermann was V.P.; and Albert Kapteyn was secretary and treasurer.

Listed below are specifications and performance data for the 3 place model S-56 as powered with the 100 h.p. Kinner K5 engine; length overall 25'0"; hite on wheels 10'1"; wing span upper 34'1"; wing span lower 30'1"; wing chord upper 67"; wing chord lower 52"; wing area upper 179 sq.ft.; wing area lower 107 sq.ft.; total wing area 286 sq.ft.; air-

The hull framework was built up of spruce and ash members that were fastened together with metal gussets and metal fittings, then covered with plywood sheet and an outer covering of fabric; the hull bottom was of double-planked cedar. The front cockpit seated two with provisions for dual controls; the rear cockpit, usually relegated as a place to put things, seated one and could be closed off with a metal panel when not in use. The wing framework was built up of spruce hollow box-beams and spruce and plywood truss-type wing ribs; the completed wing framework was covered with fabric. The simple landing gear of 64 inch tread was retracted or extended by a hand operated lever in the cockpit; wheels were 28x4 and no brakes were provided. The horizontal stabilizer was of a wooden framework and the fin, rudder, and elevators were built up of welded steel tubing; all was covered in fabric. Standard equipment included an anchor, mooring lines, life preservers, first aid kit, fire extinguisher, navigation lights, Heywood air-operated

Fig. 275 New York police dept. patrolled harbor with S-56

engine starter, and cockpit covers. For details of the improved model S-56B and S-56-31 versions as powered with the 125 h.p. Kinner B5 engine, refer to discussion in the chapter for ATC # 336.

Listed below are model S-56 entries as gleaned from registration records:

NC-192M;	Model S-56	(# 4)	Kinner K5.
NC-193N;	„	(# 5)	„
NC-194M;	„	(# 7)	„
NC-325N;	„	(# 8)	„
NC-324N;	„	(# 10)	„
NC-349N;	„	(# 12)	„
NS-371N;	„	(# 14)	„
NC-352N;	„	(# 15)	„
NC-353N;	„	(# 16)	„
NC-354N;	„	(# 17)	„
NC-355N;	„	(# 18)	„
NC-356N;	„	(# 19)	„
NC-378N;	„	(# 22)	„
NS-383N;	„	(# 23)	„
NC-382N;	„	(# 25)	„
NR-898W;	„	(# 27)	„
NC-380N;	„	(# 28)	„
NC-900V;	„	(# 30)	„
NC-901V;	„	(# 31)	„
NC-902V;	„	(# 32)	„
NC-903V;	„	(# 33)	„
NC-904V;	„	(# 34)	„
NC-905V;	„	(# 35)	„
NC-906V;	„	(# 36)	„
NC-908V;	„	(# 38)	„

Serial # 5 also with Siemens-Halske engine; registration number for serial # 6 - 9 - 11 - 20 - 21 - 24 - 26 - 29 - 37 unknown; serial # 10 - 19 - 27 - 29 later as S-56B; serial # 14-23 with N. Y. Police Dept.; serial # 27 also as S-56C; "Marchetti" airfoil used on S-56 was modified Goettingen 430.

Fig. 276 S-56 was equally at home on land or water.

Fig. 277. Eastman "Sea Rover" with Curtiss "Challenger" engine, shown high over Detroit river.

The Eastman "Sea Rover" was a small flying boat especially designed for the sportsman or for commercial operators that planned to run sight-seeing tours over lake or sea-side resorts. Being a small craft of nominal seating, Eastman never boasted needlessly that the "Sea Rover" would have any great commercial value and stated quite honestly that it was purposely designed for sport flying and just plain fun. However, to help defray expenses, some operators occasionally made a fair profit "hopping passengers" and some of these craft were used as stand-by "air ferry". Born on the broad Detroit River, a river that was ideal for craft of the "Sea Rover" type, the varied pleasures available by air in this area were boundless; many other such areas in this country were equally suited. A proposed model of the E-2 was to have been equipped with a cabin enclosure but it was never developed and a later version of the "Sea Rover" was arranged as an amphibian (E-2A) to operate off land or water.

Developed and tested early in 1928 as the Beasley-Eastman "flying boat", the prototype in this series was designed by James H. Eastman and Tom Towle; P. R. Beasley, a Detroit business-man, financed the development of this craft. First powered with an Anzani (French) engine and later with the 110 h.p.

Warner engine, the sesqui-plane prototype seated two and was quite typical to later versions except for the boom arrangement to support the tail surfaces. Production versions in 1929 were enlarged somewhat to carry three, a more powerful engine was installed and the rear portion of the hull was extended out to carry the tail-group instead of the spidery booms. Unusual in the prototyype and also in later versions, was the bracing arrangement for the wings in the form of a large vee-strut on each side, instead of the normally used criss-cross wires. With the "Sea Rover" development launched and well on its way, Tom Towle then busied himself in the design of a twin-engined amphibious monoplane and the Eastman company was soon absorbed as a subsidiary of the Detroit Aircraft Corp.

The Eastman "Sea Rover" model E-2 as shown was a good-looking flying boat of the sesqui-plane type; though actually a biplane, its very small lower wing would nearly qualify it as a "sesqui". The trim all-metal hull had ample seating for 3 but the front cockpit, which was the passenger's cockpit, did have barely enough room to accommodate a 4th person. Powered with the 6 cyl. Curtiss "Challenger" (R-600) engine of 170-185 h.p., the engine was mounted into the center-section panel of the upper wing; placed well forward to create proper distribution, the engine and its nacelle were faired neatly into the under-side of the upper wing. With adequate power reserve, the "Sea Rover" was of sprightly behavior in the air, a gross-load take off required less than 12 seconds and it was very maneuverable in the water. Because of its tractor-mounted engine, the seating seemed to be poorly placed but visibility was still quite adequate. Built of rugged and simple structure, the "Sea Rover" was quite dependable and served owners for many years. The type certificate number for the Eastman "Sea Rover" model E-2 was issued 1-22-30 and some 16 examples of this model were manufactured by the Eastman Aircraft Corp. at Detroit. Mich. Edward S. Evans was the president; James H. Eastman was chief of engineering; Carl B. Squier was the sales manager but was later sent to the Lockheed division in Calif. as general manager. In 1930 Edward S. Evans vacated the presidency to be succeeded by P. R. Beasley and Carl S. Betts became sales manager.

Listed below are specifications and performance data for the Eastman "Sea Rover" model E-2 as powered with the 185 h.p.

Curtiss "Challenger" engine; length overall 26'3"; hite overall 8'9"; wing span upper 36'0"; wing span lower 20'8"; wing chord upper 68"; wing chord lower 36"; wing area upper 190 sq.ft.; wing area lower 53 sq.ft.; total wing area 243 sq.ft.; airfoil Clark Y; wt. empty 1745 lbs.; useful load 980; payload with 48 gal. fuel was 490 lbs.; gross wt. 2725 lbs.; max. speed 110; cruising speed 90; landing speed 50; climb 740 ft. first min. at sea level; ceiling 9500 ft.; gas cap. 48 gal.; oil cap. 5 gal.; cruising range at 10.6 gal. per hour was 360 miles; price at the factory originally quoted as $8750., later raised to $9985. and drastically cut to $6750. in March of 1931.

The hull framework was built up of ash and spruce members and bulkheads that were bolted together into a robust structure of 5 water-tight compartments; outer covering was of "Alclad" aluminum alloy sheet that was screwed to the frame. Two open cockpits with seating for two in each were located under the upper wing; the pilot normally sat in the rear cockpit and a set of dual controls were available. Cockpit interiors were upholstered and each seat was fitted with a Kapok-filled leather cushion which could be used as a life-preserver. The framework of the upper wing

Fig. 278. "Sea Rover" flying boat offered boundless pleasures in lake-studded areas.

Fig. 279. Beasley-Eastman was prototype with Anzani engine; differed considerably from production version.

was built up of spruce and plywood box-type spar beams with spruce and plywood truss-type wing ribs; the center-section panel which mounted the engine, was built up of welded chrome-moly steel tubing. The lower wings were built up of solid spruce spar beams with spruce and plywood truss-type wing ribs; all leading edges were covered with dural sheet and the completed framework of both wings was covered in fabric. Upper and lower wing panels were connected together by two N-type interplane struts and braced to the hull by two large vee-type struts which eliminated the normal wire bracing. Small floats of metal-covered wooden frames were mounted near the wing tips of lower wing to keep wings clear of the water. The two fuel tanks were mounted in the center-section panel of the upper wing and the oil tank was mounted in the engine nacelle. The fabric covered tail-group was built up of welded steel tubing and steel channel sections; the horizontal surfaces were mounted high on the fin to clear water and the stabilizer was adjustable in flight. A water-rudder was inter-connected to the flight rudder by a spring-loaded attachment and all controls were operated by stranded steel cable. A metal propeller, air-operated Heywood engine starter, navigation and mooring lights, fire extinguisher, and a first-aid kit were

Fig. 280. Test landing of "Sea Rover" on snow-covered ground.

standard equipment. The next development of the "Sea Rover" was the model E-2A amphibian as described in the chapter for ATC # 338.

Listed below are Eastman "Sea Rover" model E-2 entries as gleaned from registration records:

X-3643; Beasley-Eastman (# 1)
                    Anzani & Warner 110.
  -3019; Sea Rover E-2   (# 2)
                    Curtiss Challenger.
X-592M;   „   (# 3)   „
NC-460M;   „   (# 4)   „
NC-461M;   „   (# 5)   „
NC-462M;   „   (# 6)   „
NC-463M;   „   (# 7)   „

NC-474M;   „   (# 8)   „
NC-465M;   „   (# 9)   „
NC-464M;   „   (# 10)   „
NC-466M;   „   (# 11)   „
NC-467M;   „   (# 12)   „
NC-468M;   „   (# 13)   „
NC-469M;   „   (# 14)   „
NC-470M;   „   (# 15)   „
NC-471M;   „   (# 16)   „
NC-472M;   „   (# 17)   „
NC-473M;   „   (# 18)   „

Serial # 3 was used as prototype for E-2A amphibian; serial # 10 later as E-2D with Packard Diesel engine; serial # 11 and # 15 later converted to E-2A amphibian.

A.T.C. #289
(1-14-30)
FAIRCHILD, MODEL 71-A

Fig. 281. Fairchild 71-A with "Wasp" engine; sweep-back of wing visible in ground shadow.

Introduced in 1928 as a refinement of the FC-2W2, the Fairchild model 71 was a culmination of all the features and requirements that were dictated by many years of service experience - an experience accumulated on the rugged frontiers of aviation in many different parts of the world, where "Fairchild" monoplanes had been serving these past few years. The "Seventy One" was a versatile airplane adaptable to many chores and it served profitably on many of the early air-lines in this country; especially popular in our neighbouring countries its dependable service in various parts of our western hemisphere earned it the reputation as "queen of the bush country". Gentle, rugged and trustworthy, a good many of the "Model 71" led long and useful lives; at least 20 operated constantly for some 10 years and some were still flying in the late fifties.

As pictured here, the Fairchild model 71-A was a strut-braced high wing cabin monoplane of buxom proportion with ample seating for 7; interior arrangements and appointments were improved slightly over the earlier "Seventy One" of 1928 and empty weight was increased by several hundred pounds but otherwise it was typical. That is,

it was basically typical except for the 5 degree of sweep-back that was built into the wing. Technical descriptions of the 71-A never divulge the reason why sweep-back in the wing was necessary but we can safely assume that it was either to create a better weight distribution around the C. G. (center of gravity) or perhaps to improve the airplane's directional stability. This version of the 71 is an extremely rare model and only a few have been recorded in registration records. One of this models was with Braniff and American Airways, one was with Alaskan Airways and one served as a test-bed for the Bell Telephone Laboratories. The type certificate number for the Fairchild model 71-A was issued 1-14-30 and at least 3 examples of this model were manufactured by the Fairchild Airplane Mfg. Corp. at Farmingdale, Long Island, N. Y. Sherman M. Fairchild was president; and Ralph C. Lockwood was the chief engineer.

Listed below are specifications and performance data for the Fairchild model 71-A as powered with the 420 h.p. Pratt & Whitney "Wasp" engine; length overall 32'11"; hite overall 9'4"; wing span 50'0"; wing chord 84"; total wing area 332 sq.ft.; airfoil Goettingen 387; wt. empty 3156 lbs.; useful load 2344;

payload with 148 gal. fuel was 1200 lbs.; gross wt. 5500 lbs.; max. speed 135; cruising speed 110; landing speed 57; climb 900 ft. first min. at sea level; climb in 10 min. was 6200 ft.; ceiling 15,300 ft.; gas cap. 148 gal.; oil cap. 12 gal.; cruising range at 22 gal. per hour was 700 miles; price at the factory field was $18,900. drastically reduced to $13,500. in May of 1931.

The construction details and general arrangement of the Fairchild 71-A were quite typical to the model 71 as described in the chapter for ATC # 89 of U.S. CIVIL AIR-CRAFT, Vol. 1 in addition to the following. Cabin dimensions were 3 ft. wide x 9 ft. long x 54 in. high and all seating could be quickly removed for hauling cargo; a baggage compartment of 145 cu. ft. capacity with allowance for 217 lbs. was aft of the cabin. The landing gear of 90 inch tread used oleo-spring shock absorbing struts; wheels were 32x8 and Bendix brakes were standard equipment. Two fuel tanks of 55 gal. capacity each were mounted in the root ends of each wing half and a 40 gal. tank was mounted in the cabin roof. The leading edges of the wing were covered with mahogany plywood and the completed framework was covered in fabric. Engine nose-cowling had controllable shutters to regulate the temperature in the engine section. A metal

propeller, carburetor-air heater, cabin heater, navigation lights, hand-crank inertia-type engine starter, and exhaust tail-pipe were standard equipment. The next Fairchild development was the model 51 as described in the chapter for ATC # 357.

Listed below are Fairchild model 71-A entries as gleaned from registration records:
NC-9154 ; Model 71-A (# 638) Wasp 420.
NC-9170 ;   ,,   (# 642)   ,,
NC-952V;   ,,   (# 3501)   ,,

Serial # 3501. on Group 2 approval numbered 2-290 as 5 place special.

It is amazing to note that agricultural experts had finally come to the conclusion that airplanes were responsible for the spreading of weeds, a source of added worries to the poor farmer. The sudden appearance of strange weeds in a locality far removed from where such pests had been thriving, had brought about an investigation which disclosed that weed seeds had been clinging to airplane tires and were being transported to various other areas. It was recommended that quarantine officers were to be stationed at landing fields to intercept the weed seeds and keep the seeds from doing their dirty work.

Fig. 282. Fairchild 71 leaving factory field on delivery to some far-off customer.

Fig. 283. Spartan C3-166 with Comet 7-E engine; C3-166 was last of Spartan biplane series.

Since the first "Spartan" biplane was introduced in 1926 by Mid-Continent, this 3 place open cockpit craft of the general-purpose type had been continuously improved in detail and was brought out in at least 9 different versions. Always an advocate of the radial air-cooled engine, Willis C. Brown, the designer of the original "Spartan" series, had not much choice in this type of engine during 1925-26; with hopes of doing better later, he chose to use the 9 cyl. "Super Rhone" (a static radial that was converted from the LeRhone "rotary" of World War 1) in the prototype airplane. For production versions of the C-3, Brown had to look to engines of foreign make and finally chose the 9 cyl. Ryan-Siemens engine of 125 h.p. for the model C3-1. When labor strife in Germany posed to halt his supply of the Siemens-Halske engines, Brown hopefully tested the new 4 cyl. Fairchild-Caminez engine of 120-130 h.p. but it was far from being satisfactory. In some disgust and desperation, Brown again had to turn to foreign engines and this time selected the 9 cyl. Walter engine of 120-130 h.p. from Czecho-slovakia for the model C3-2. Later in 1928

suitable American engines were beginning to hit the market and several were tested by Spartan in rapid succession. The 6 cyl. Curtiss "Challenger" engine of 170 h.p. was mounted in the model C3-3 but it was felt that this would not be the right combination for the mild-mannered "Spartan". Next the 7 cyl. Axelson engine of 115-150 h.p. was installed in the model C3-4 but no great enthusiasm was provoked by this combination either. For the model C3-5 the new 5 cyl. Wright J6 engine of 150-165 h.p. was anxiously tried and the "Spartan" seemed to have fallen into the right combination; as a production version, the C3-5 was modified slightly into the C3-165 that was approved on ATC # 195. Anxious to fill the demands for craft of still higher performance, the "Spartan" was offered as the model C3-225 with the 7 cyl. Wright J6 engine of 225 h.p. To offer a suitable running-mate to the popular C3-165, "Spartan" mounted and tested the new 7 cyl. "Comet" 7-E engine of 165 h.p. in the model C3-166 as the final version in this biplane series.

The handsome "Spartan" model C3-166 as

pictured here, was an open cockpit biplane with seating for three and was also of the general-purpose type. Planned as a running-mate for the other two versions in the series, the model C3-166 was typical in most all respects except for its powerplant installation which was the "Comet" 7-E of 165 h.p. With performance and behavior well comparable to the model C3-165, the model C3-166 was an interesting development that was apparently nipped in the bud by uncertain economic conditions and a drastic curtailment in air-craft sales. Tested satisfactorily and finally approved, the type certificate number for the "Spartan" model C3-166 as powered with the 165 h.p. Comet 7-E engine was issued 1-20-30 and only one example of this model was manufactured by the Spartan Aircraft Co. at Tulsa, Okla. Lawrence V. Kerber was president; Rex B. Beisel was V.P. and chief engineer; L. R. Dooley was sales manager and Wilton M. Briney, former sales manager at Butler Aircraft, was also on the sales staff; R. W. Grigsby was chief pilot in charge of test and development.

Listed below are specifications and performance data for the "Spartan" model C3-166 as powered with the "Comet" 7-E engine; length overall 23'5"; hite overall 8'10"; wing span upper & lower 32'0"; wing chord both 60"; wing area upper 151 sq.ft.; wing area lower 140 sq.ft.; total wing area 291 sq.ft.; airfoil Clark Y; wt. empty 1637 lbs.; useful load 968; payload with 45 gal. fuel was 483 lbs.; payload with 65 gal. fuel was 363 lbs.; gross wt. 2605 lbs.; max. speed 118; cruising speed 100; landing speed 47; climb 800 ft. first min. at sea level; climb in 10 min. was 6950 ft.; ceiling 12,000 ft.; gas cap. normal 45 gal.; gas cap. max. 65 gal.; oil cap. 6.5 gal.; cruising range at 10 gal. per hour was 450-600 miles; price at the factory field was $5675.

Construction details and general arrangement for the "Spartan" model C3-166 was typical to those versions as described in the chapters for ATC # 195 and # 286 of U.S. CIVIL AIRCRAFT, Vols. 2 and 3. The only modifications apparent in the model C3-166 were those necessary for the installation of the "Comet" engine. Along with the "Spartan" biplanes that were displayed at the annual Detroit Air Show for 1930, were a new line of "Spartan" cabin monoplanes that had been in development for the past year. The first of these shown was the C4-225, a tidy high winged cabin monoplane for four that was a concession to the trend which was shaping for the future. The model C4-225 as described in the chapter for ATC # 310 was powered with the "Whirlwind Seven" and bore typical "Spartan" lines.

Listed below is information on the only known example of the model C3-166: NC-707N; Model C3-166 (# 151) Comet 7-E.

Designation for this model was first as C3-165C, later changed to C3-166; according to registration records, this craft was later converted to C3-165 specifications by installation of a Wright R-540 engine.

Fig. 284. Consolidated "Fleetster" model 17 with "Hornet" engine; shown here in prototype.

The flashing "Fleetster" model 17 was certainly a revolutionary departure from the conventional types as normally built by Consolidated Aircraft; designed as a small air-liner for shuttle-service on lines in difficult territory, it is at least commendable that the model 17 should be patterned after the most exciting airplane of this period. One need not stretch the imagination too strongly to notice that the configuration of the cigar-shaped "Fleetster" was no doubt inspired and influenced in this direction by the famous Lockheed "Vega". The end results of this selection of form were very gratifying and Consolidated was very proud of their achievement. Well versed in management and certainly adept in securing the business for his company, Rueben Fleet was largely instrumental in the forming of the N.Y.R.B.A. Line to link the two Americas by air. With Consolidated already working on a large order of "Commodore" flying boats for the N.Y.R.B.A. Line to be used in coast-hugging flights to So. America, the "Fleetster" was especially designed for service with this same line on routes in the interior of Argentina and

neighbouring countries. Soon, "Fleetster" flights were penetrating the Argentine interior over pampas, impenetrable jungle, and vast mountain peaks in trail-blazing fast mail and passenger service. The N.Y.R.B.A. Line had a route from Buenos Aires to Santiago, Chile which lie across the Andes; another route radiated from Buenos Aires into Bolivia and another into Paraguay and Uraguay. The South Americans loved air-travel and in many cases made the transition from burro to airplane like a duck going to water. The N.Y.R.B.A. Line had 3 "Fleetster" model 17 in service and at least one operated on twin-float seaplane gear; in a deal consummated 9-15-30 Pan American Airways System acquired these 3 "Fleetster" along with 10 "Commodore" flying boats. With routes now greatly extended both the "Commodore" and the "Fleetster" remained in Pan American service.

The "Fleetster" model 17 as pictured here, was a trim and clean high winged cabin monoplane with a seating arrangement for 8 in the circular all-metal monocoque fuselage. The thick pure cantilever wing was of wood construction which required no external bracing and when equipped with "tear drop" wheel fairings, the "Fleetster" was almost a pure streamlined form; it was able to sustain a top speed of 180 m.p.h. with ease. The "Fleetster" was also equipped with the N.A.C.A. low-drag engine cowling and was one of the few to use this type of engine fairing at this time; beset with a few problems at first on some installations, this type of engine streamlining was used almost universally a few years later. Powered with the 9 cyl. Pratt & Whitney "Hornet" B engine of 575 h.p. which was encased in an N.A.C.A. cowling, the "Fleetster" had an impressive performance with power reserve yet available for those extra demands. Three of the "Fleetster" model 17 were put into service for rapid transit of medium loads on the N.Y.R.B.A. Line and at least one of them was equipped with seaplane gear. Another example in this same series became the personal air-taxi of F. Trubee Davison, Asst. Sec. of War; his pilot on cross-country inspection tours was the incomparable Capt. Ira C. Eaker. The type certificate number for the "Fleetster" model 17 both as landplane and seaplane was issued 1-23-30 and 4 examples of this model were manufactured by the Consolidated Aircraft Corp. at Buffalo New York. Rueben Hollis Fleet was president and general manager; Lawrence D. "Larry" Bell was V.P.; and I. M. Laddon, designer of the "Fleetster", was chief engineer for the heavy aircraft division.

Listed below are specifications and performance data for the Consolidated "Fleet-ster" model 17 as powered with the 575 h.p. "Hornet" B (R-1860) engine; length overall 31'9"; hite overall 9'2"; wing span 45'0"; wing chord at root 106"; wing chord at tip 68"; total wing area 313.5 sq.ft.; airfoil Goettingen 398 mod.; wt. empty 3326 lbs.; useful load 1974; payload with 145 gal. fuel was 844 lbs.; gross wt. 5300 lbs.; max. speed (cowling & pants) 180; cruising speed 153; landing speed 60; climb 1200 ft. first min. at sea level; climb to 10,000 ft. was 13.5 min.; ceiling 18,000 ft.; gas cap. 145 gal.; oil cap. 12 gal.; cruising range at 32 gal. per hour was 675 miles; price at the factory was $27,500. lowered to $26,500. later in 1931. Landplane certificate later amended to allow 5600 lbs. gross wt.; some of this increase was used for extra equipment and balance was used for extra payload; performance changes were as follows; landing speed 65; climb 1050 ft.; ceiling 17,000 ft. The following figures are for model 17 mounted on Edo J twin-float seaplane gear; wt. empty 3822 lbs.; useful load 1748; payload with 145 gal. fuel was 818 lbs.; gross wt. 5570 lbs.; max. speed 170; cruising speed 145; landing speed 60; climb 1050 ft. first min. at sea level; ceiling 17,000 ft.; cruising range at 32 gal. per hour was 600 miles.

The fuselage was a metal monococque structure of circular section that was built up of duralumin annular rings and bulkheads, longitudinal stringers and stiffeners to which was riveted the smooth aluminum alloy outer skin. The main cabin for the seating of six was arranged with two full-width seats that accommodated 3 across; arrangement was

Fig. 285. "Fleetster" 17 on floats; served So. America for N.Y.R.B.A. line.

available in the normal method where all seats faced forward or in a "chummy" arrangement where the front seat faced backwards and the rear seat faced forward. There was a 25 cu. ft. baggage compartment to the rear which was not normally used with a full passenger load. When payload allowance would permit, a canvas bag could be suspended in back of the rear seat for 30 lbs. of baggage. The pilot's cockpit up forward had ample room for the pilot and one passenger but this seat was eliminated when carrying radio gear. The main cabin section of 62x80x54 inch dimension could be stripped of all seating for carrying large loads of cargo. The cantilever wing framework was built up of two heavy box-type spar beams and two auxiliary spar beams of spruce flanges and plywood webs; spruce and plywood compression members and truss-type former ribs completed the structure which was covered with plywood sheet. Fuel tanks were mounted in the center portion of the wing. The split-axle landing gear of 96 inch tread was of the long-leg type and used oleo-spring shock absorbing struts; wheels were 32x6 and Bendix brakes were standard equipment. The tail wheel was mounted deep in the fuselage and was covered with a streamlined metal boot. Long individual exhaust stacks were normally used for better performance but a collector-ring was available. The all-metal

tail-group was built up of duralumin box spars and channel forming ribs covered with a riveted aluminum alloy skin; the metal skin covering was stiffened with shallow corrugations. Of unsymmetrical shape to offset engine torque, the fin was built integral with the fuselage and the horizontal stabilizer was adjustable in flight. A metal propeller, navigation lights, electric inertia-type engine starter, generator, and battery, were standard equipment. Optional equipment included streamlined wheel fairings (wheel pants), radio gear and shielding, landing lights, parachute flares, special leather upholstery, extra-long exhaust tail-pipes, and 35x15 low-pressure airwheels. The next "Fleetster" development was the parasol-winged model 20 as described in the chapter for ATC # 320.

Listed below are "Fleetster" model 17 entries as gleaned from various records:

NC-657M; Model 17-1 (# 1) Hornet B.
NC-671M;   „   (# 2)   „
NC-700V;   „   (# 3)   „
NC-672M;   „   (# 4)   „

Serial # 1-2-4 originally with N.Y.R.B.A., bought 9-15-30 by Pan Am; serial # 3 modified as 17 Special for F. Trubee Davison, later modified as 6 place model 17-2 on Group 2 approval numbered 2-219; serial # 2 and # 4 junked by Pan Am on Oct.-1934.

Fig. 286. "Fleetster" 17 special for personal use of F. Trubee Davison.

A.T.C. #292
(1-23-30)
STEARMAN "JUNIOR SPEEDMAIL", 4-E

Fig. 287. Stearman "Junior Speedmail" model 4-E with 420 H.P. "Wasp" engine.

As a progressive development and further refinement of the basic "Stearman" design, the 4-series biplane was a high-performance open sport plane that was also easily adaptable to various commercial uses. Somewhat bigger and bulkier than the standard model C3R but not nearly as big as the earlier "Speedmail", the new "Junior Speedmail" was an in-between model that was probably the handsomest airplane that Stearman ever built. Fitted with the bulbuous N.A.C.A. low-drag engine cowling as an integral part of the configuration, the "Stearman 4" was the first production biplane to utilize this deep engine fairing with any measure of success; the resulting performance caused other heretofore skeptical manufacturers to take a new look at the advantages to be gained with this type of air-cooled engine streamlining. As a wait and see policy, the new series "Stearman" was formally introducd to the flying public as the model 4-E with the 420 h.p. Pratt & Whitney "Wasp" engine, specifically as a thundering and very colorful high performance airplane for the sportsman-pilot or for the business-house planning to do promotion that would catch the public eye. With the series soon well proven and firmly entrenched in the minds of those in the industry, it was but a small matter to modify the basic design slightly to adapt it to specific needs of private-owner and air-line customers. As a consequence, the company fared well with the "Stearman 4", which was finally produced in 3 different models and each of these models was convertible to carrying air-mail and cargo, plus yet a few versions that were built for special purpose. The flawless design, careful workmanship, and built-in safety that had been characteristic of the "Stearman" airplane in the past, seemed to be even more outstanding and more pronounced in the make-up of the 4-series.

As pictured here in several views, the "Stearman" model 4-E was an open cockpit biplane of beautiful proportion that was carefully engineered around the big N.A.C.A. engine cowling; though seemingly large for the seating of three, the large cross-section of the oval fuselage was necessary to fair out the large diameter up front. Powered with the 9 cyl. "Wasp" C1 engine of 420 h.p. or the SC1 of 450 h.p., the model 4-E had a fairly outstanding performance that translated into a dashing personality. Though primarily arranged as a sport-craft for 3, with allowance for plenty of baggage and extra equipment, this payload capacity of more than 600 lbs. was also readily adapted to the hauling of mail and cargo. With outward appearance changed considerably, the aerodynamic arrangement of the model 4-E was still quite typical to the basic design laid out a few years back, so flight characteristics and general behavior were more or less comparable to those that had made the "Stearman" biplane a perennial favorite. The rugged structure and inherent reliability built into the "Stearman 4" naturally promoted longevity so it is not surprising that at least 20 examples of various models in this series were still in active service some 8 or 9 years later. The type certificate number for the "Stearman" model 4-E was issued 1-23-30 and some 11 examples of this model were manufactured by the Stearman Aircraft Co. at Wichita, Kansas. Frederick Rentschler of the United Aircraft & Transport Corp. was chairman of the board.

Listed below are specifications and performance data for the "Stearman" model 4-E as powered with the 420 h.p. "Wasp" engine; length overall 26'4"; hite overall 10'2"; wing span upper 38'0"; wing span lower 28'0"; wing chord upper 66"; wing chord lower 51"; wing area upper 204 sq.ft.; wing area lower 103 sq.ft.; total wing area 307 sq.ft.; airfoil Goettingen 436; wt. empty 2426 lbs.; useful load 1510; payload with 106 gal. fuel was 629 lbs.; gross wt. 3936 lbs.; max. speed 158: cruising speed 130; landing speed 55; climb 1400 ft. first min. at sea level; ceiling 18,000 ft.; gas cap. 106 gal.; oil cap. 10 gal.; cruising range at 23 gal. per hour was 580 miles; price at the factory field was $16,000.

The fuselage framework was built up of welded chrome-moly steel tubing, that was extensively faired to an oval shape with wooden formers and fairing strips, then fabric covered. The cockpits were deep and well protected with a door on the left side for entry to the front cockpit; there was a small baggage compartment of 5 cu. ft. capacity with allowance for 40 lbs. just in back of the rear cockpit and a larger baggage compartment forward of the front cockpit with allowance for 249 lbs. The wing framework was built up of solid spruce spar beams that were routed out for lightness with spruce and plywood Warren truss-type wing ribs; the leading edges were covered with dural sheet and the completed framework was covered with fabric. The Friese type balanced-hinge ailerons were in the upper wing and were actuated by an external push-pull strut of streamlined steel tubing, connected to rods, tubes, and bellcranks in the lower wing. The large N.A.C.A. engine cowling was removable in sections for servicing and the lower wing was neatly faired into the fuselage junction. The main fuel tank of 62 gal. capacity was mounted high in the fuselage just ahead of the front cockpit and

Fig. 288. Ideal for sportsmen, this 4-E owned by Hollywood films director.

extra fuel was carried in a 44 gal. tank mounted in the center-section panel of the upper wing. The split-axle landing gear of 96 inch tread was of the out-rigger type and uesd "Aerol" shock absorbing struts; wheels were Kelsey-Hayes or Bendix in 30x5 or 32x6 with brakes as standard equipment. The swiveling tail-wheel was mounted in the very end of the fuselage with a notch in the rudder to provide working clearance. The fabric covered tail-group was built up of welded chrome-moly steel tubing; the fin was ground adjustable and the horizontal stabilizer was adjustable in flight. A Hamilton-Standard metal propeller, hand-crank inertia-type engine starter, navigation lights, and wheel brakes were standard equipment. Dual controls, a duplicate instrument panel in front cockpit, parachute flares, radio receiver, battery, generator, electric engine starter, landing lights, streamlined wheel fairings, and 9.50x12 semi-airwheels were offered as optional equipment. The next "Stearman" development in the 4-series was the 300 h.p. model 4-C as described in the chapter for ATC # 304.

Listed below are Stearman model 4-E entries as gleaned from registration records:

NS-2; Model 4-E (# 4001) Wasp 420.
NC-664K;   „   (# 4004)   „

Fig. 289. Stearman 4-E offered high speed and substantial pay-load.

NC-663K;	„	(# 4005)	„
NC-666K;	„	(# 4006)	„
NC-667K;	„	(# 4007)	„
NC-674K;	„	(# 4008)	„
NC-770H;	„	(# 4010)	„
CF-AMB;	„	(# 4016)	„
CF-AMC;	„	(# 4017)	„
NC-779H;	„	(# 4019)	„
NC-791H;	„	(# 4023)	„

Serial # 4001-4006-4007 first as model 4-C on Group 2 approval 2-155; serial # 4005 as 4-E equipped with wheel pants; serial # 4010 later as 4-CM and later as 4-EM (CF-ASF) to Canada; serial # 4016-4017 later as 4-EM to Canada; serial # 4019 first as model 4-D; serial # 4020 as 4-EX on Group 2 approval 2-279; serial # 4021-4022 as 4-E Special on Group 2 approval 2-278.

Fig. 290. Stearman 4-E had lines of beauty, often mentioned as finest "Stearman" ever built.

Fig. 291. American Eagle model 201 with 100 H.P. Kinner K5 engine.

The "Model 201" was part of the new concept in "American Eagle" all-purpose bi-planes as planned for 1930. Where previous models were all the same basic air-frame with but various engines mounted to suit different uses, the new model 201 was designed specifically for the Kinner K5 engine and in a configuration that was arranged especially to do the job that would be required in a lighter and smaller air-frame with the immediate gain of better performance and somewhat better utility. Evidence shows a trend was slowly forming to break away from the one basic design altered only by horse-power and to offer now combinations of engine and airplane that were thoughtfully designed for a price range or a particular usage. A half-hearted trial in this direction was introduced earlier by American Eagle in the "Phaeton" series (Model 251) where the spread of horsepower was only 60 but due to peculiararities of this design it would not hardly allow a power differential of more than that. So, to offer a new low powered general-purpose airplane that would take the place of the previous model A-129, American Eagle started from scratch and designed

the model 201, a sensible craft that was carefully planned to be the most efficient in daily service with the average private-owner or the small flying-service operator. American Eagle now had an average of some 200 employees working in the shops about this time and they were all no doubt thankful for the Model 201 because it was the only thing that they had selling to any extent in 1930. To stimulate the clearance of some left-over model A-129 that were still on hand, American Eagle offered to throw in a $1500. flight training course free for the price of the airplane alone. The offer didn't create any great land-slide of eager customers but it did help to clear surplus stocks and make way for the up-coming new models.

Because the model 201 was designed specifically to do only certain jobs within limitations and without making allowances in its make-up for installations of extra power, it was thereby lighter, smaller, and more efficient. Because it was designed to operate effectively on approximately 100 h.p., there was still allowance for a power reserve to give it very good performance and a behavior that was not bogged down by excesses of any kind. To prove its sprightly nature and capability, young Mildred Kauffman, a talented aviatrix, flew a model 201 through 46 easy loops to set a new record for women flyers to shoot at. Of course the new model 201 was not quite as good as American Eagle was trying to make everyone believe by its fantastic claims for this airplane in their advertisements but nevertheless it was a very good airplane. One thing is certain, had this model been offered in better times it is quite likely that it would have sold in the hundreds. Staying pretty well in the same power range, the 201 was tested also with the 110 h.p. Warner engine, the 90 h.p. Brownback "Tiger", and the 90 h.p. inverted Cirrus Ensign but no production was scheduled for any of these versions because it was decided to standardize with the 100 h.p. Kinner.

The "American Eagle" model 201 was a trim open cockpit biplane with seating for three; designed for general-purpose work at the low-budget level, it was carefully arranged to be as efficient as possible with sensible use of the power available. Many described it as better than the earlier "American Eagle" A-129 (ATC # 124) and that is testimony enough that it was a pretty good airplane. However, due to its entirely different aerodynamic and mechanical arrangement, it was an airplane of unlike personality that was hardly no kin to the gentle and kind A-129. In contrast, the 201 was like a sprightly tomboy inclined to be playful at times but never unruly; altogether it can be summed up that

the 201 had good reason for its popularity and was one of the best airplanes that American Eagle ever built. Despite the fact that the mortality rate for this type of airplane was rather high, at least 18 of the 44 built were still in active service some 10 years later. The type certificate number for the Model 201 as powered with the Kinner K5 engine, was issued 1-28-30 and some 44 examples of this model were manufactured by the American Eagle Aircraft Corp. at Fairfax Field in Kansas City, Kansas. Ed. Porterfield, Jr. was president; Jack E. Foster was chief engineer; Larry D. Ruch was chief pilot; and D. H. Hollowell, formerly V.P. in charge of sales, was replaced by John Carroll Cone recently with Command-Aire. Of considerable interest is the fact that serial # 801 of this series had a double distinction; it was the first of 44 examples built of the "Model 201" and quite a few years later, after American Eagle as a company was already gone, this craft was modified into the first and only "Porterfield" biplane that was built by Ed Porterfield as a sport-trainer.

Listed below are specifications and performance data for the "American Eagle" model 201 as powered with the 100 h.p. Kinner K5 engine; length overall 22'11"; hite overall 8'5"; wing span upper 31'1"; wing span lower 28'8"; wing chord both 57"; wing area upper 133 sq.ft.; wing area lower 120 sq.ft.; total wing area 253 sq.ft.; airfoil USA-35B; wt. empty 1168 lbs.; useful load 782; payload with 36 gal. fuel was 370 lbs.; gross wt. 1950 lbs.; max. speed 115; cruising speed

Fig. 292. Model 201 was popular for its good nature and high performance.

*Fig. 293. Model 201 was new concept in
American Eagle all-purpose biplanes;
remained favorite for many years.*

95; landing speed 38; climb 680 ft. first. min.
at sea level; ceiling 14,000 ft.; gas cap. 36 gal.;
oil cap. 4 gal.; cruising range at 6.5 gal. per
hour was 475 miles; price at the factory field
was $3995. with price holding up throughout
production period.

The fuselage framework was built up of
welded chrome-moly steel tubing, heavily
faired to a streamlined shape with wooden
fairing strips and fabric covered. The cockpits were deep, roomy, and well protected
with the front cockpit a full 33½ in. wide;
entry to the front cockpit was by a large
door on left side which was an easy step from
the wing-walk. The fuel tank was mounted
high in the fuselage just ahead of the front
cockpit with a large direct-reading fuel gauge
projecting thru the cowling. The wing framework was built up of solid spruce spar beams
that were routed out for lightness with
spruce and mahogany plywood built-up wing
ribs; the leading edges were covered to preserve the airfoil form and the completed
framework was covered with fabric. The
split-axle landing gear of 65 inch tread was
of the cross-axle type with "oleo" shock
absorbing struts; wheels were 26x4 and no
brakes were provided. Low pressure airwheels
were available after May 1930; tail skid was
of the spring-leaf type with a removable
shoe. The fabric covered tail-group was built
up of welded steel tubing with fin being
ground adjustable to compensate for engine
torque and the horizontal stabilizer was adjustable in flight. The noisy Kinner was well
muffled by a large volume nose-type collector
ring and tail-pipe. A wooden Supreme propeller was standard equipment. A metal propeller, airwheels, and Heywood air-operated
engine starter were optional. It might be
noted that the "Model 201" was the last of

the "American Eagle" biplanes; the next
development was the D-430 monoplane as
described in the chapter for ATC # 301.

Listed below are "American Eagle" model
201 entries as gleaned from registration
records:

NC-582H;	Model 201	(# 801)	Kinner K5.
NC-216N;	„	(# 802)	„
NC-217N;	„	(# 803)	„
NC-218N;	„	(# 804)	„
NC-219N;	„	(# 805)	„
NC-220N;	„	(# 806)	„
NC-222N;	„	(# 807)	„
NC-223N;	„	(# 808)	„
NC-224N;	„	(# 809)	„
NC-225N;	„	(# 810)	„
NC-226N;	„	(# 811)	„
NC-227N;	„	(# 812)	„
NC-229N;	„	(# 813)	„
NC-230N;	„	(# 814)	„
NC-232N;	„	(# 815)	„
NC-233N;	„	(# 816)	„
NC-235N;	„	(# 817)	„
NC-239N;	„	(# 818)	„
NC-234N;	„	(# 819)	„
NC-240N;	„	(# 820)	„
NC-241N;	„	(# 821)	„
NC-237N;	„	(# 822)	„
NC-242N;	„	(# 823)	„
NC-244N;	„	(# 824)	„
NC-243N;	„	(# 825)	„
NC-249N;	„	(# 826)	„
;	„	(# 827)	„
NC-246N;	„	(# 828)	„
NC-268N;	„	(# 829)	„
NC-271N;	„	(# 830)	„
;	„	(# 831)	„
NC-270N;	„	(# 832)	„
NC-273N;	„	(# 833)	„
;	„	(# 834)	„
NC-274N;	„	(# 835)	„
NC-282N;	„	(# 836)	„
NC-283N;	„	(# 837)	„
NC-289N;	„	(# 838)	„
NC-288N;	„	(# 839)	„
NC-290N;	„	(# 840)	„
NC-292N;	„	(# 841)	„
NC-450V;	„	(# 842)	„
NC-463V;	„	(# 843)	„
NC-464V;	„	(# 844)	„

Registration numbers for serial # 827-831-
834 unknown.

Fig. 294. Stinson SM-8B with 225 H.P. Wright J6 engine was 1930 version of "Junior" series.

The popular "Junior" started out as the "baby" of the Stinson monoplane line, a typical craft in scaled-down fashion that seated either 3 or 4 with a fairly good performance on a meager 110 h.p.; it was originally Eddie Stinson's intent to offer this light cabin monoplane as a personal type airplane that would handle a substantial payload with economy for the private owner or the small business man. The "Detroiter Junior" as introduced in 1928 (ATC # 48) was not exactly a dainty little sprite to begin with and successive power increases in the interests of better performance, as demanded by prospective customers, caused the craft to become increasingly bigger and much heavier. It was not too long before this baby of the Stinson line grew itself right out of the light-plane class. In progressive developments to follow, the horsepower mounted in the "Junior" was raised to 165, to 175, to 220, and finally to 225 in the model SM-2AC; it needs only a little imagination to visualize that the "baby Detroiter" as it started out was hardly the same breed of airplane that was coming off the line in late 1929. Not wishing to deviate too far from a good thing, the proven SM-2 design was modified slightly and improved

for 1930 only in minor detail and introduced as the SM-8 series. Stinson Aircraft had enjoyed a good year in 1929; some 120 airplanes were built and sold and when the nation's economy began to look scarish in the closing months of 1929, the company was merged with the huge Cord Corp. to strengthen its financial structure. Despite the dark outlook created by the business-depression and the many laments voiced by others in the industry, Stinson busily added to their plant area and was getting ready for a big year in 1930.

The Stinson "Junior" model SM-8B as discussed here, was a high winged cabin monoplane of rather buxom proportion that had ample seating for 4, with an outward arrangement and appearance that remained fairly typical to the earlier SM-2AC. The various changes not easily seen at first glance were incorporated into the new design in the interests of increased comfort, a bit more practical utility and a structure more conducive to efficient manufacture. Features planned to improve the airplane yet hold a price range that would be attractive enough to stimulate buying. Powered with the 7 cyl. Wright J6 (R-760) engine of 225 h.p. the new

Fig. 295. Models of new "Junior" series were typical except for engine installations.

SM-8B seemed to show somewhat less performance than comparable models in the SM-2 series, so, it was either because of some changes in general design, which does seem to be apparent or else the performance figures released were more honest and true to the actual fact. Especially leveled at the business man and quite suitable as a family-type airplane the model SM-8B was seen occasionally in many different parts of the country; the Dept. of Commerce had several in use on their "airways division" and the Mars Candy Co. (maker of the famous "Mars" candy bars) used one of this type in business promotion. Priced fairly high because of the relatively high cost of Wright engines, the model SM-8B sold rather slow and a price reduction of some $2000. was not enough to create any volume buying; because of this the SM-8B was a rare and rather scarce model in the new series. The type certificate number for the Stinson "Junior" model SM-8B was issued 2-1-30 and some 5 or more examples of this model were manufactured by the Stinson Aircraft Corp. at Wayne, Mich.; a division of the Cord Corp. with offices in Chicago, Ill.

Listed below are specifications and performance data for the Stinson "Junior" model SM-8B as powered with the 225 h.p. Wright J6 engine; length overall 29'0"; hite overall 8'9"; wing span 41'8"; wing chord 75"; total wing area 234 sq.ft.; airfoil Clark Y; wt. empty 2063 lbs.; useful load 1137; payload with 65 gal. fuel was 542 lbs.; gross wt. 3200 lbs.; max. speed 128; cruising speed 108; landing speed 50; climb 800 ft. first min. at sea level; ceiling 14,000 ft.; gas cap. 65 gal.; oil cap. 5 gal.; cruising range at 12.5

gal. per hour was 540 miles; price at the factory field was $10,500. early in 1930, later lowered to $8495.

The fuselage framework was built up of welded chrome-moly steel tubing and mild-carbon steel tubing, deeply faired to shape with wooden and metal formers, wooden fairing strips and fabric covered. The cabin interior was neat and tastefully upholstered in broadcloth fabrics and all cabin windows were of shatter-proof glass; the front side panels at the pilot's station could be lowered or raised. A sky-light in the forward part of the cabin roof provided vision upward and a large door on each side provided easy entry to the cabin. The wing framework was built up of heavy-sectioned solid spruce spars that were routed to an I-beam section, with spruce and plywood truss-type wing ribs; the leading edges were covered with dural sheet and the completed framework was covered in fabric. The out-rigger landing gear of 115 inch tread used oleo-spring shock absorbing struts; wheels were Kelsey-Hayes 30x5 and brakes were standard equipment. The fabric covered tail group was built up of welded chrome-moly and mild-carbon steel tubing; the fin was ground adjustable and the horizontal stabilizer was adjustable in flight. A metal propeller, wiring for navigation lights, tail wheel, electric inertia-type engine starter, and cabin heater were standard equipment. The next development in the Stinson "Junior" monoplane was the model SM-8A as described in the chapter for ATC # 295 in this volume.

Listed below are SM-8B "Junior" entries as gleaned from registration records:
NC-419M; Junior SM-8B (# 4010)

			Wright R-760.
NS-27;	„	(# 4204)	„
NS-26;	„	(# 4206)	„
NS-28;	„	(# 4210)	„
NC-424Y;	„	(# 4257)	„

Serial # 4001 was also an SM-8B but there was no listing in registration records; serial # 4204-4206-4210 with Dept. of Commerce; serial # 4257 with Mars Candy Co.

Fig. 296. Stinson "Junior" model SM8A with 210 H.P. Lycoming engine; shown here on factory field.

At a time when other manufacturers were reducing their working forces and turning to other economy measures just to stay in business, Stinson Aircraft added to their plant capacity and boldly laid plans for a big year ahead. In a merger with the huge Cord Corp., Stinson found itself in an enviable position with large sums of capital at their disposal, tremendous buying power, and a source for good engines within their own corporation to power their new model. For the year 1930 the previous "Junior" design was modified and improved slightly to become the basis for the SM-8 series; powered with the newly developed Lycoming engine of 210 h.p. the model SM-8A was groomed for a market that had temporarily sagged but needed only the promise of good value to revive some interest. With pride and fanfare the "Junior" model SM-8A, as powered with the new 9 cyl. "Lycoming" engine, was announced in April of 1930 with the unbelieveable price-tag of $5775.; a $10,000. airplane for this amount of money caused everyone to sit up and take notice. This was certainly establishing a brand new standard of value within the industry. Though the price of the SM-8A "Junior" was certainly not justified on early sales volume, it was based on the belief that this airplane finally would sell in a terrific volume to justify the low price. As it turned out their calcultations had been right. Thirty

two of the **SM-8A** were sold in April of 1930, 51 were sold in the month of May and the production schedule for June called for at least 75 airplanes; literally hundreds of these new "Juniors" were flying in all parts of the country before long. By year's end the model SM-8A was sold in a number greater than all other cabin airplanes put together; there was nothing on the market that compared with this airplane in price or in value.

The Stinson "Junior" model SM-8A was a high winged cabin monoplane with seating for four in an arrangement that was quite comfortable and well appointed; one would think that an airplane with this price-tag would surely have to sacrifice comfort and interior finery for the sake of economy in production, but this was certainly not the case with the SM-8A. A sound-proofed cabin with shatter-proof glass windows had yet the added features of cabin heat and ventilation plus pilot aids such as an electric engine starter, wheel brakes, and an emergency brake for parking; these were the same features as those found in any $10,000. airplane. Powered with the 9 cyl. Lycoming (R-680) engine of 210-215 h.p. the SM-8A delivered very good performance with a utility of proven value. With the new "Junior" definitely slated for the business-man or for the private owner-flyer who would enjoy owning such an air

Fig. 297. SM-8A extremely popular as business-plane; ship shown was flying show-case for Bendix products.

plane but had no time or inclination to developing a superb piloting technique, it had to be designed as a safe and easy to fly airplane, an airplane that would literally pull itself out of tight spots and allow more than the average margin for error in most any instance. The enviable safety record and the interesting lore left behind by the Stinson "Junior" series is good and lasting proof that this airplane was every bit as good as it was planned and perhaps even better. It is known fact in flying circles everywhere that "Stinson stories" are a legend and favored with such a special sanctity that if you were caught in the middle of a tall airplane story that sounded somewhat incredible, you need but say that you were flying a Stinson SM- or SR- something or other and your reputation for honesty would go unmarred! The type certificate number for the Stinson "Junior" model SM-8A was issued 2-11-30 for landplane and this approval was later amended to include the seaplane on Edo P twin-float gear. It would be quite a task to give a correct and full tally but we do know that several hundred of the model SM-8A were

built by the Stinson Aircraft Corp. at Wayne, Mich.

Listed below are specifications and performance data for the Stinson "Junior" model SM-8A as powered with the 215 h.p. Lycoming engine; length overall 28'11"; hite overall 8'9"; wing span 41'8"; wing chord 75"; total wing area 234 sq.ft.; airfoil Clark Y; wt. empty 2061 lbs.; useful load 1134; payload with 61 gal. fuel was 564 lbs.; gross wt. 3195 lbs.; max. speed 125; cruising speed 105; stall speed 55; landing speed 50; climb 780 ft. first min. at sea level; ceiling 14,000 ft.; gas cap. 61 gal.; oil cap. 5 gal.; cruising range at 12 gal. per hour was 500 miles; price at the factory field was $5775. in April of 1930, raised to $5995. in Nov. of 1930. The following figures are for seaplane as mounted on Edo P pontoon gear; wt. empty 2440 lbs.; useful load 1080; payload with 60 gal. fuel was 515 lbs.; gross wt. 3520 lbs.; max. speed 115; cruising speed 95; landing speed 60; climb 650 ft. first min. at sea level; ceiling 12,000 ft.; cruising range at 12 gal. per hour was 450 miles; price at factory was near $8000.

The construction details and general arrangement for the model SM-8A are typical to those of the SM-8B as described in the previous chapter, including the following features explained in more detail. The cabin area was of 40x54x48 inch dimension and 3 of the seats could be easily removed for hauling cargo; a baggage compartment of

Fig. 298. SM-8A during tests as twin-float seaplane.

Fig. 299. The SM-8A on floats was popular in Canada.

7 cu. ft. capacity with allowance for 50 lbs. was to the rear of the cabin section with access from the outside. A cabin heater was provided and front side-panel windows rolled down for ventilation; an emergency brake was provided for parking. The wing panels, in two halves, had fuel tanks mounted in the root end of each wing; ailerons were of metal construction with a large steel tube spar and sheet steel ribs. The wing bracing struts were of large diameter steel tubes that were encased in a balsa-wood fairing; these fairings were shaped to an "Eiffel 380" airfoil section to provide added lift and help promote lateral stability. The rugged landing gear used oil-spring shock absorbing struts of Stinson manufacture and the wheel brakes were also of Stinson make. The tail wheel was moved further up from the tail-post, thereby eliminating the rudder cut-out as used on earlier "Junior" models. Entry doors to the cabin were quite large and a novel step was provided on each side to make boarding the airplane an easy task. An adjustable metal propeller and an electric engine starter were also standard equipment. Extra equipment available included an engine speed-ring cowling, streamlined wheel fairings (wheel

Fig. 300. 1931 version of SM-8A "Junior" sported speed-ring cowling and low-pressure tires.

pants), skis for winter flying, and 8.50x10 semi-balloon tires and wheels. The next development in the Stinson "Junior" series was the model SM-7A as discussed in the chapter for ATC # 298 in this volume.

Listed below are only the first few SM-8A entries as gleaned from registration records:

NC-409M; Junior SM-8A (# 4002)
Lycoming R-680.

NC-411M;	,,	(# 4003)	,,
NC-209W;	,,	(# 4004)	,,
NC-416M;	,,	(# 4005)	,,
NC-415M;	,,	(# 4006)	,,
NC-417M;	,,	(# 4007)	,,
NC-213W;	,,	(# 4008)	,,
NC-418M;	,,	(# 4009)	,,
NC-420M;	,,	(# 4011)	,,
NC-421M;	,,	(# 4012)	,,
NC-422M;	,,	(# 4013)	,,
NC-425M;	,,	(# 4014)	,,
NC-423M;	,,	(# 4015)	,,
NC-208W;	,,	(# 4016)	,,
NC-1026;	,,	(# 4017)	,,
NC-203W;	,,	(# 4018)	,,
NC-904W;	,,	(# 4019)	,,
NC-429M;	,,	(# 4020)	,,
NC-426M;	,,	(# 4021)	,,
NC-218W;	,,	(# 4022)	,,
NC-427M;	,,	(# 4023)	,,
NC-215W;	,,	(# 4024)	,,
NC-207W;	,,	(# 4025)	,,
NC-428M;	,,	(# 4026)	,,
NC-212W;	,,	(# 4027)	,,
NC-242W;	,,	(# 4028)	,,
NC-205W;	,,	(# 4029)	,,
NC-224W;	,,	(# 4030)	,,
X-424M;	,,	(# 4100)	,,

Serial # 4014 later as SM-8A Special on Group 2 approval 2-461 with J5 engine; serial # 4100 was first seaplane.

A.T.C. #296
(2-12-30)
FORD "TRI-MOTOR", 5-AT-CS

Fig. 301. Unusual setting for the Ford "Tri-Motor"; shown is model 5-AT-CS.

The famous Ford "Tri-Motor", affection-ately known as the "Tin Goose", was certainly a well known form to most everyone who had any interest in airplanes at all but several rarely seen versions of this familiar design did appear from time to time; the model 5-AT-CS "seaplane" was one that not too many got to see. Easily adapted from the popular model 5-AT-C land-plane, the 5-AT-CS seaplane was only slightly modified and equipped for test with model A-27000 twin "floats" of Aircraft Products manufacture. First flown in Sept. of 1929 by Leroy Manning, Ford's versatile test pilot, the huge 5-AT-CS seaplane was tested and demonstrated for a time from the broad and calm Detroit River, creating quite a local sensation. It was still quite unusual at this time to see such a large craft operating from the water and interested on-lookers flocked eagerly to see each flight. Some time later the 5-AT-CS was tested and evaluated at the Philadelphia Navy Yard as a torpedo-bomber; whether tests were satis-factory to the U.S. Navy or not is not known but no orders were placed for any craft of this type. Another 5-AT-CS seaplane was tested by TWA air-lines in an operation from the harbor near down-town New York; this was an experiment in operation to offer local commuter service. One other example is be-lieved to have operated as a seaplane on floats in Chile for a short time. There was certainly

nothing unusual about an airplane on floats, there were hundreds in operation all over the country but an airplane the size of the Ford tri-motor on floats was very much unusual and the lessons learned with this experiment led to float installations on yet bigger airplanes.

The Ford "Tri-Motor" model 5-AT-CS was a high winged cabin monoplane of the 5-AT-C type (ATC # 165) and was arranged for the seating of 13 passengers with a pilot and co-pilot. No reports were available to indicate as to its behavior or character but perform-ance figures seem to bear out that the "Ford" was not greatly hampered by the two large floats. The large pontoons and their neces-sary attachments accounted for a 1200 lb. increase in the empty weight over the 5-AT-C landplane, so consequently the useful load and payload were cut by that amount. Though bracing of the float gear to the airplane seems somewhat complicated, it is probable that any of the model 5-AT-C could have been converted to a seaplane of this type without any extensive modifications. Power for the 5-AT-CS version was three Pratt & Whitney "Wasp" C1 engines of 420 h.p. each or three SC1 engines of 450 h.p. each; outboard engines were usually fitted with Townend type "speed-ring" cowlings. The type certificate number for the Ford-Stout model 5-AT-CS

Fig. 302. 5-AT-CS shown here at factory.

as powered with three 450 h.p. "Wasp" engines was issued 2-12-30 and at least 3 examples of this model were manufactured by the Stout Metal Airplane Co., a division of the Ford Motor Co. at Dearborn, Michigan.

Listed below are specifications and performance data for the Ford "Tri-Motor" seaplane model 5-AT-CS as powered with three 450 h.p. "Wasp" engines; length overall 51'4"; hite overall 14'6"; wing span 77'10"; wing chord at root 156"; wing chord at tip 92"; total wing area 835 sq.ft.; airfoil Goettingen 386; wt. empty 8675 lbs.; useful load 4825; payload with 277 gal. fuel was 2595 lbs.; payload with 355 gal. fuel was 2095 lbs.; gross wt. 13,500 lbs.; max. speed 130; cruising speed 104; landing speed 64; climb 900 ft. first min. at sea level; climb in 10 min. was 6800 ft.; ceiling 14,200 ft.; gas cap. normal 277 gal.; gas cap. max. 355 gal.; oil cap. 30-34 gal.; cruising range at 66 gal. per hour

Fig. 303. 5-AT-CS shown during tests at Philadelphia Navy Yard.

was 400-500 miles; price at the factory was $68,000. soon lowered to $64,000.

The construction details and general arrangement of the model 5-AT-CS seaplane was typical to the 5-AT-C landplane as described in the chapter for ATC # 165 of U.S. CIVIL AIRCRAFT, Vol. 2, in addition to the following. Because of the higher empty weight imposed by the bulky twin-float gear, the 5-AT-CS interior arrangement was limited to 13 passengers and a crew of two. A baggage compartment of 30 cu. ft. capacity had allowance for 550 lbs.; the tread or center-line of the float gear was

14 ft. 9 in. For this conversion the tail wheel was removed and area was extended onto lower end of the normal flight rudder. The next development in the Ford tri-motor series was the model 9-A as described in the chapter for ATC # 307.

Listed below are 5-AT-CS entries as gleaned from registration records :
NC-410H; 5-AT-CS (# 5-AT-69) 3 Wasp 450.
NC-414H;   „   (# 5-AT-74)   „
NC-415H;   „   (# 5-AT-75)   „

This certificate for serial # 5-AT-74 only.

Fig. 304. Float-equipped "Ford" tested in New York for commuter service.

Fig. 305. Mohawk "Pinto" M-1-CW with 110 H.P. Warner engine; ideal for the sportsman.

A direct descendant of the earlier "Spur-wing", the Mohawk model M-1-CW was offered as a companion model to the M-1-CK "Pinto" that was discussed here in a previous chapter (ATC # 263). Also as mentioned before, the new design was relieved of much of its capricious habits that had been told and re-told in sessions around the hangar until they had reached a proportion far out of line from the whole truth. The early "Pinto" in its close-coupled configuration was a playful airplane indeed and chances are that it would tend to get the upper hand when guided by an inexperienced or timid pilot but somewhat like a young bronco, it could be tamed. The new series "Pinto" as redesigned by John Akerman was a craft of a much more placid nature but there was still plenty of eager verve and playful nature left in its make-up to make each flight a rewarding experience. Definately leveled at the sportsman-pilot or the private owner who flew just for the fun of it, the new "Pinto" posed as an airplane that should have become quite popular; that it did not was just a quirk of fate and prevailing circumstances.

Along with the development of the

"Spurwing", Mohawk Aircraft had two more models in test that were slated for possible manufacture in the future. One was the "Redskin" which was a low-winged mono-plane of typical "Pinto" lines but the open cockpits were faired over by a streamlined canopy that was dubbed a coupe cabin for three; it is doubtful if there was room enough for 2 occupants in either one of the cockpits. The other model in test was the M-2-C, a small twin-engined airplane of basically similar configuration but stretched out somewhat into a bigger craft with seating ample for three. The small "twin" appeared of good sound design and was powered with two Michigan "Rover" engines of 55-60 h.p. each, anyone of which could have flown the airplane and maintained level flight if need be. Both the enclosed "Redskin" and the light cabin "twin" held promise of being popular additions to the Mohawk line but financial difficulties prevented further development of these two designs. A "Super Pinto" with the 165 h.p. Wright J6 engine was also planned but never developed. Mohawk tried for a foreign market but several other American aircraft manufacturers had the same idea and there was not enough business to go around. A pioneer in the low-wing design, Mohawk Aircraft had tough going throughout its career and finally had to give in to the economic depression of the early "thirties"; Rufus R. Rand, Jr. tried to hold the company together long enough to ride out the economic storm but the "storm" lasted too long.

The Mohawk "Pinto" model M-1-CW was an open cockpit low-winged monoplane of handsome proportion that seated two in tandem in an arrangement that provided excellent visibility and easy entry or exit to the cockpits. Powered with the 7 cyl. Warner "Scarab" engine of 110 h.p. the "Pinto" was quite lively and had power reserve to execute just about any of the advanced aerobatic maneuvers; there is no doubt that it would have been an excellent trainer for the secondary phases. Basically designed as a sport-plane its lean and hardy character was of rugged frame that stood up well to abuse and hard usage. Had conditions been more favorable it is quite likely that the M-1-C "Pinto" would surely have been built in much greater number. The type certificate number for the Mohawk "Pinto" model M-1-CW, as powered with the 110 h.p. Warner engine, was issued 2-15-30 and some 3 or more examples of this model were manufactured by the Mohawk Aircraft Corp. at Minneapolis, Minn. When John D. Akerman took over to redesign the early "Pinto", Wallace "Chet" Cummings, designer of the original series, left

Fig. 306. Mohawk "Spurwing" was prototype for new series.

for Wichita to design and develop the Watkins "Skylark". Akerman, formerly with the Hamilton Metal Airplane Co., left Mohawk late in 1930 to become professor of aeronautics at the Milwaukee University. Rufus R. Rand, Jr. was appointed receiver of the failing company but by 1931, Mohawk Aircraft was on the way to being but a memory.

Listed below are specifications and performance data for the Mohawk "Pinto" model M-1-CW as powered with the 110 h.p. Warner engine; length overall 24'2"; hite overall 7'7"; wing span 34'11"; wing chord at root 75"; wing chord at tip 34"; wing chord mean 53"; total wing area 145 sq.ft.; airfoil Mohawk #3; wt. empty 1124 lbs.; useful load 658; payload with 32 gal. fuel was 269 lbs.; gross wt. 1782 lbs.; max. speed 118; cruising speed 100; landing speed 40; climb 900 ft. first min. at sea level; ceiling 15,500 ft.; gas cap. 32 gal.; oil cap. 4 gal.; cruising range at 6.5 gal. per hour was 475 miles; price at the factory first announced as $6000., soon reduced to $5100. and later reduced to

Fig. 307. "Spurwing" fitted with coupe-type canopy was "Redskin."

$4995.

The fuselage framework was a long narrow backbone built up of welded chrome-moly steel tubing, deeply faired to shape with wooden formers and fairing strips then fabric covered. The cockpits were deep and well protected from the elements by large windshields; a fairly large door provided easy entry to each cockpit. Dual joy-stick controls were provided with the front stick quickly removable when not in use. The fuel tank was mounted high in the fuselage just ahead of the front cockpit and a baggage compartment of 9.3 cu. ft. capacity was in the turtleback section behind the rear cockpit. The cantilever wing framework in two halves, was built up of spruce flanges and mahogany plywood webs into box-type spar beams, with wing ribs of spruce trusses around mahogany plywood webs; the leading edges were covered with plywood sheet to the front spar, the Friese-type ailerons were of a steel tube structure and the completed framework was covered with fabric. The split-axle landing gear of 102 inch tread used "Aerol" (air-oil) shock absorbing struts; wheels were 26x4 and

Bendix brakes were standard equipment. The landing gear attachment and arrangement was still similar to that on the early "Pinto", except that the oleo-legs were now on the inside of the wheels to permit easier removal and better servicing. The fabric covered tail-group was built up of welded chrome-moly steel tubing; the fin was ground adjustable and the horizontal stabilizer was adjustable in flight. A Hartzell wooden propeller, wiring for navigation lights, a spring-leaf tail skid, fire extinguisher, log books, tool kit, first-aid kit, engine and cockpit covers were standard equipment. A metal propeller, navigation lights, and Heywood air-operated engine starter were available as extra equipment.

Listed below are "Pinto" M-1-CW entries as gleaned from registration records:
NC-555E; Pinto M-1-CW (# 1) Warner 110.
NC-556E;   „   (# 2)   „
NC-180N;   „   (# 4)   „

Serial # 1 was first as "Redskin", later modified to M-1-CW; serial # 2 first as "Spurwing", later modified to M-1-CW; serial # 4 first as M-1-CK.

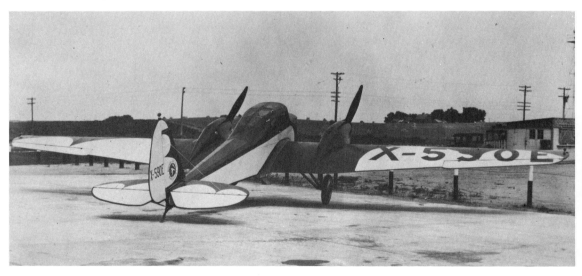

Fig. 308. Mohawk M-2-C was experimental light "twin."

Fig. 309. Stinson "Junior" model SM-7A with 300 H.P. Wright J6 engine.

In the new line-up for 1930, Stinson Aircraft planned to concentrate more on the "Junior" type with offerings in two different series; the SM-7 series in two versions were of higher horsepower and the SM-8 series in three versions were of lower power and more modest price-tags. Knowing from past experience that certain customers would be interested in an airplane of the 4 place cabin type but with a personality containing much more verve and capable of a much better performance, Stinson singled out the model SM-7 "Junior" as the offering for this type of clientele. We might say that the model SM-7 was the "Junior" version designed for the sportsman-pilot or the business-man who wouldn't hesitate paying a little more for a much better airplane. Since 1926 Stinson airplanes had roamed the frontiers of aviation and had earned a reputation for performance and dependability that was almost a household word among the people that fly. Though this was quite flattering to the people involved in the creation of these airplanes, such a reputation loomed often to impose increasing problems. The problem now being that every time a new model or new series was designed and introduced to the market, the flying public had to be shown and duly convinced not only that it was a better airplane but that it was just as good as the one before. Like it or not, we know that this was rather hard to

do at times and for Stinson this would be increasingly harder to do as the years went by.

The Stinson "Junior" model SM-7A as discussed here was a high winged cabin monoplane that seated 4 in a style and comfort that was equal to the best that anyone had to offer. Of a proportion and arrangement quite typical to the SM-8 series, the SM-7A was offered as a craft with allowable gross weight that would permit more fuel for a greater cruising range and yet have useful load enough to allow the addition of any number of accessories such as speed-ring engine cowling, wheel streamlines, and perhaps even a radio, without cutting into the payload too much. Powered with the 9 cyl. Wright J6 (R-975) engine of 300 h.p. there was plenty of performance available with flight characteristics and general behavior of a nature that had made the "Junior" a country-wide favorite for many years. There were airplanes that were certainly more handsome, airplanes with more speed and airplanes that excelled in one way or another but there were few airplanes that could match the "Junior" as an all-round good airplane. The type certificate number for the Stinson "Junior" model SM-7A, as powered with the 300 h.p. Wright J6 engine, was issued 2-18-30 and some 8 or more examples of this model were manufactured by the Stinson Aircraft Corp. at

Wayne, Mich.; a division of the Cord Corp. E. L. Cord was chairman of the board; Edward A. Stinson was president; Wm. A. Mara was V.P.; B. D. DeWeese was V.P. and general manager; Kenneth M. Ronan and Wm. C. Naylor were project engineers; and A. H. Saxon became chief engineer later in the year. With a plant capacity now of some 100,000 sq.ft., more than 400 craftsmen and women were soon employed and production potential had risen to nearly 20 airplanes per week.

Listed below are specifications and performance data for the Stinson "Junior" model SM-7A as powered with the 300 h.p. Wright J6 engine; length overall 30'11"; hite overall 8'9"; wing span 41'8"; wing chord 75"; total wing area 234 sq.ft.; airfoil Clark Y; wt. empty 2234 lbs.; useful load 1266; payload with 90 gal. fuel was 506 lbs.; gross wt. 3500 lbs.; max. speed 142; cruising speed 120; landing speed 60; climb 1000 ft. first min. at sea level; ceiling 18,000 ft.; gas cap. 90 gal.; oil cap. 7 gal.; cruising range at 18 gal. per hour was 550 miles; price at the factory field was $10,495., later reduced to $8995.

The construction details and general arrangement for the model SM-7A were typical to the models SM-8B and SM-8A as described in the chapters for ATC # 294 and # 295 in this volume, including the following. Because of its higher price the model SM-7A was upholstered in richer fabrics and the interior was bedecked with accessories that were only available as extra equipment in the SM-8 series. For that extra in performance the SM-7A was available with a low drag "speed ring" engine cowling and wheel streamlines (wheel pants) that boosted the top speed to some 150 m.p.h. An adjustable metal propeller, electric inertia-type engine starter, engine exhaust muffler, wheel brakes, emergency parking brake, tail wheel, wiring for night-flying equipment, 7 cu. ft. baggage compartment with allowance for 50 lbs., dual wheel controls, and cabin heater were standard equipment. A battery, generator, skis for winter flying, navigation and landing lights, parachute flares, and mountings for radio in the baggage compartment were available as extra equipment. Another development in the Stinson "Junior" series was the model SM-8D powered by the Packard "Diesel" as described in the chapter for ATC # 312.

Listed below are "Junior" model SM-7A entries as gleaned from registration records: NC-410M; Junior SM-7A (# 3000)

			Wright R-975.
NC-414M;	„	(# 3001)	„
NC-216W;	„	(# 3002)	„
NC-222W;	„	(# 3004)	„
NC-248W;	„	(# 3005)	„
NC-406Y;	„	(# 3008)	„
NC-980W;	„	(# 3011)	„
NC-499Y;	„	(# 3100)	„

Serial # 3100 listed also as SM-7AS which may either denote seaplane or special version.

Fig. 310. SM-7A was high-performance version of new "Junior" series.

A.T.C. #299
(3-12-30)
BACH "AIR YACHT", 3-CT-9S

Fig. 311. Model 3-CT-9S was plush version of Bach "Air Yacht."

With the utility and high performance that was built into the model 3-CT-9 "Air Yacht", it was only logical that Bach would want to offer this craft in a plush version suitable for use as an executive-transport or perhaps as a "flying yacht" for some play-boy millionaire. To attract the eye of this type of clientele the model 3-CT-9S was custom-finished as to exterior, in addition to the custom interior. Though quite adaptable to most any desired arrangement, a typical interior contained a 4 piece over-stuffed set which included a full-length divan or couch and 3 comfortable chairs. To complement this motif of comfort was a combination cabinet that contained a fold-out table, an ice-box, humidors for cigars and cigarettes, a caraffe, and drinking glasses. The floor was naturally covered with a deep plush rug and fancy curtains adorned the windows. Towards the rear of the cabin section was a compact and complete lavatory which included a mirror and a dressing table. The cabin walls were sound-proofed, insu-lated, and paneled in a pattern of rich-looking polished wood veneer; it was almost impossible to muffle the thundering roar of the three big engines but conversation was possible at slightly above normal levels. In addition to these deluxe surroundings, the model 3-CT-9S had inherent utility that was possible by its effortless high performance; ports of call need not be mammoth airports and lofty terrain certainly was no problem, the Bach had ability to master it all. There is no doubt that the 3-CT-9S was tailor-made for the purpose planned but "executives" on the average were not rushing out to buy $40,000. airplanes and the ranks of the "play-boy millionaire" were thinning rapidly during this particular period.

The Bach "Air Yacht" model 3-CT-9S was a tri-motored high wing cabin monoplane of a medium size that was typical to the pre-viously described model 3-CT-9 (ATC # 271) except for its custom exterior finish and de-luxe interior appointments. Quoted seating capacity for this model was nine but this does not seem likely in view of the space that would be taken up by all the extra equipment and the 183 lb. loss in useful load. Powered with one Pratt & Whitney "Wasp" engine of 450 h.p. in the nose and two Wright J6 (R-760) engines of 225 h.p. each in the nacelles, this total of 900 h.p. was mainly the secret to the exceptional performance that could be expected of the 3-CT-9S. A proposed development similar to the 3-CT-9S but lighter and of slightly better performance, was planned to be powered with 3 Wright "Whirlwind Nine" engines of 300 h.p. each; this would have been a complete break-away from the normal Bach custom of using mixed power combinations but the development apparently never went beyond the planning stage. The model 3-CT-9S was first certificated on a Group 2 approval numbered 2-179 but this was superseded by ATC # 299. The type certificate number for the Bach "Air Yacht" model 3-CT-9S was issued 3-12-30 and records available show that only one example was built; it can be readily assumed that modifications eligible for this certificate could easily be performed to any of the existing model 3-CT-9. The Bach Aircraft Co., Inc. was located on the Los Angeles Metropolitan Airport in Van Nuys, Calif. T. Warden Hunter of Detroit was now the president, succeeding B. L. Graves; L. Morton Bach was V.P. in charge of engineering; C. W. Faucett was secretary-treasurer; Waldo D. Waterman and W. J. "Pat" Fleming performed most of the test-flying chores.

Listed below are specifications and performance data for the Bach "Air Yacht" model 3-CT-9S; length overall 36'10"; hite overall 9'9"; wing span 58'5"; wing chord at root 132"; wing chord at tip 96"; total wing area 490 sq.ft.; airfoil "geometrically tapered Clark Y"; wt. empty 5193 lbs.; useful load 2807; payload with 195 gal. fuel was 1340 lbs.; gross wt. 8000 lbs.; max. speed 162; cruising speed 136; landing speed 60; climb 1380 ft. first min. at sea level; ceiling 20,190 ft.; gas cap. 195 gal.; oil cap. 16 gal.; cruising range at 42 gal. per hour was 600 miles; price at the factory field was $40,000. and up.

Construction details and general arrangement for the model 3-CT-9S were typical to the model 3-CT-9 as described in the chapter for ATC # 271, except for deluxe interior appointments as described in first paragraph. An interesting accessory were the "fenders" that were mounted over each wheel to deflect stones and debris from the spinning wheels away from the whirling propellers; these fenders were sometimes replaced by conventional streamlined wheel fairings. The main cabin area was 46 in. wide x 11 ft. long x 60 in. high; adaptable for all sorts of interior arrangements or stripped of all seating for the hauling of cargo. The pilots cockpit was 40 in. x 40 in. and the crew on the 3-CT-9S would normally be a pilot and also a co-pilot that usually would be required to act as steward. Typical to many other aircraft of this period, the wing bracing struts on the "Air Yacht" served double duty; shaped to an Eiffel 376 airfoil section of 20 in. maximum chord, the struts contributed some 47 sq. ft. of lift which when added to the normal wing area would be some 537 sq. ft. of supporting surface. Then too, the angle of the struts acted as exaggerated dihedral and contributed measurably to the lateral stability.

Beset by problems that were instigated by the economic depression, production schedules had to be cut drastically by the middle of 1930; further development had more or less ceased and only a few of the Bach "tri-motors" were built during this shaky period. Bach however did build a model 3-CT-9K in 1931 that was powered with a "Wasp" 450 in the nose and two Kinner C-5 engines of 210 h.p. each in the wing nacelles. Of the remaining examples that were built in the previous 2 years, several plied the air-ways in the west for a few years more and some were exported for service in our neighbouring countries. At least one of the tri-motors was modified into a single-engined freighter and was used to "dust" bananas in Central America. Manufacturing operations gamely continued with a skeleton crew for a time but by the end of 1931, Bach Aircraft was all but gone.

Listed below is the only known example of the model 3-CT-9S as gleaned from registration records:

NC-317V; Bach 3-CT-9S (# 20) Wasp 450 & 2 Wright R-760.

Fig. 312. Prototype "Sirius" model 8 built for Chas. A. Lindbergh.

Designed by Gerry Vultee from earlier designs of the low-winged "Explorer" as laid out by Jack Northrop, the new Lockheed "Sirius" gained almost instant fame for itself because the first customer for this new craft was the world's most famous flyer — one Chas. A. Lindbergh. A chance meeting between Vultee, Carl B. Squier, and "Lindy" at the National Air Races for 1929, brought forth requirements of a new airplane the intrepid flyer (Lindbergh) had in mind and was anxious to have built; pencilled notes and a rough diagram were later translated to the drawing board and the well known "Sirius" was the result. First introduced for test as an open cockpit monoplane, diminutive Anne Lindbergh, the Colonel's almost inseparable wife, suggested a sliding enclosure over the airy cockpits for added comfort; before the Lindberghs accepted delivery of their "Sirius", the enclosure was installed and this feature was then added to all examples of this model as standard equipment. After watching his ship progress through every step of its production and later test-flying the gleaming creation to his satisfaction, Col. Lindbergh and wife Anne flew from Glendale, Calif. to New York City in 14 hours and 45 minutes

to set a new west-east record. For a period after, the Lindberghs used their well-loved "Sirius" as personal transportation to points around the country and on business trips in connection with "Lindy's" work as a consultant for two of the largest air-lines. In search of increased performance for some specialized flying, Lindbergh had a 575 h.p. Wright "Cyclone" engine installed in the "Sirius" and then put it on twin-float seaplane gear; this combination was used on survey flights that later reached to nearly every corner of the world.

The beautiful low-winged "Sirius" fired the flying fraternity with anxious enthusiasm and before even the first example was finished, the company was busy working on a half dozen or so more. The second example to leave the line was bought by the Shell Oil Co. and used on promotion flights by John A. MacCready; on one short jaunt the capable Major set an unofficial speed record for commercial airplanes at 206 m.p.h. A steep dive and a high-speed pass across the airport during an air-show caused a terrific tail flutter in the "Sirius" and MacCready was very well shaken at the experience. Nearly as fast

Fig. 313. Lockheed "Sirius" during early tests.

as the "Sirius" monoplanes were being completed and delivered, record attempts were being planned to various corners of the earth. Mexican pilot Roberto Fierro in the "Anahuac" made the first non-stop flight from New York to Mexico City; another was to be used on a New York to Paris flight but a crack-up caused a cancellation and the ship was rebuilt for Chas. Kingsford-Smith as the "Lady Southern Cross". Another "Sirius" named "Justice for Hungary" made the trans-Atlantic flight from New York to Hungary with pilots Magyar and Endres; a flight from Cuba to Spain was made by another a few

years later. In the meantime, other ships of the "Sirius" clan stayed close to home and became fun and personal transportation for several sportsman-pilots; some others were converted with large cargo holds up front to haul air-mail and air-express cargo.

The Lockheed "Sirius" model 8 was a low-winged cantilever monoplane seating 2 in tandem and it was basically designed for long-range flying; easily adaptable to modification, the payload could be arranged to include large amounts of baggage for extensive cross-country trips or the fuselage hold could

Fig. 314. Lockheed "Explorer" set pattern for "Sirius"; Harold Bromley shown during tests.

be converted to carry cargo. In comparison with the model 8, the model 8-A was basically similar except for more fin and rudder area and a slight difference in the weights; the 8-A being 82 lbs. lighter when empty and there was a corresponding increase in payload. In a version called the "Sport Cabin Sirius" (model 8-C), a cabin for two was provided up forward in addition to the two occupants seated under the sliding enclosure. Quite versatile and quite easy to modify for special purposes the "Sirius" was later fitted with a retractable landing gear and this version became the model 8-D or the famous "Altair", which in turn was redesigned to become the forerunner of the speedy "Orion", the world's fastest air-liner. Powered with the 9 cyl. "Wasp" engine of 450 h.p., the "Sirius" model 8 was a wee-bit less sprightly than the "Wasp-Vega" and several hundred pounds heavier when empty but otherwise its performance was more or less on a par with others in the Lockheed line-up. Fitted with the N.A.C.A. low-drag engine cowling as standard equipment, the "Sirius" was also available with tear-drop wheel fairings and this combination was good for nearly 180 m.p.h. Flight characteristics and general behavior were very satisfactory and the ability of this design to utilize more horsepower than the high-wing Vega eventually led to the design of faster airplanes based on this configuration. The type certificate number for the Lockheed "Sirius" models 8 and 8-A as powered with the 420-450 h.p. "Wasp" engine was issued 3-14-30 and some 14 or more examples of this model were manufactured by the Lockheed Aircraft Corp. at Burbank, Calif.; a divison of the Detroit Aircraft Corp. After the tragic loss of veteran Herb Fahy due to an accident, test-pilot duties were taken over by Marshall "Babe" Headle.

Listed below are specifications and performance data for the Lockheed "Sirius" models 8 and 8-A as powered with the 420 h.p. "Wasp" engine; length overall 27'6" (27'10"); hite overall 9'0" (9'3"); wing span 42'10"; wing chord at root 102"; wing chord at tip 63"; total wing area 294 sq.ft.; airfoil at root Clark Y-18; airfoil at tip Clark Y-9.5; wt. empty 3060 (2978) lbs.; useful load 1540 (1622); payload with 150 gal. fuel was 400 (477) lbs.; gross wt. 4600 lbs.; max. speed 175; cruising speed 150; landing speed 60; climb 1200 ft. first min. at sea level; ceiling

Fig. 315. "Sirius" flown by Capt. Macready for Shell Oil Co.

20,000 ft.: gas cap. normal 150 gal.; oil cap. 10-15 gal.; cruising range at 22 gal. per hour was 975 miles; price at the factory field was $18,985. with standard equipment. Figures in brackets are for model 8-A.

The fuselage framework of the "Sirius" was of wooden monococque construction and quite similar to that of the "Vega" or the "Air Express"; the only difference were the cockpit cut-outs and the wing attach cut-out on the bottom side. The cockpits were deep, roomy, and well protected from the weather by a sliding canopy enclosure; because most of the "Sirius" were used for extra-long jaunts, dash panels were well stocked with all the latest instruments for aerial navigation. The main fuel tank was mounted in the fuselage forward and extra fuel was carried in two tanks that were mounted in the wing; a 50 cu. ft. baggage compartment with allowance for up to 200 lbs. was in the mid-section of the fuselage. This compartment could also be modified to 75 cu. ft. capacity for a 400 lb. payload or a cabin could be arranged for the seating of two. The cantilever wing framework was built up of box-type spar beams and girder-type wing ribs; the completed framework was covered with plywood veneer. The outer plywood surfaces of the wing and the fuselage were given an added covering of fabric to add strength and provide a base for a better finish. The split-axle landing gear of 144 inch tread used oleo shock absorbing struts; wheels were 32x6 and Bendix brakes were standard equipment. The cantilever tail-group was built up of a wooden construction very similar to the wing; the horizontall stabilizer was adjustable in flight, and all movable surfaces were statically balanced. A Hamilton-

Fig. 316. Lindbergh accepting delivery on new "Sirius"; canopy over cockpits suggested by Mrs. Lindbergh.

Standard metal propeller, an Eclipse inertia-type engine starter, navigation lights, cockpit lights, battery, fire extinguisher, and N.A.C.A. engine cowling were standard equipment. Among optional equipment available for the "Sirius" were dual controls, wheel fairings, tail wheel, extra fuel tanks, and optional colors and color schemes. The next Lockheed development was the metal-fuselage "Vega" model DL-1 as described in the chapter for ATC # 308.

Listed below are Lockheed "Sirius" entries as gleaned from various records:

NR-211;	Sirius 8	(# 140)	Wasp 420-450.
NC-349V;	„ 8	(# 141)	„
NR-12W;	„ 8	(# 142)	„
NC-13W;	„ 8-A	(# 143)	„
NC-14W;	„ 8	(# 144)	„
NC-15W;	„ 8-A	(# 145)	„
NC-16W;	„ 8-A	(# 146)	„
XB-ADA;	„ 8	(# 149)	„
NR-116W;	„ 8-C	(# 150)	„
NC-117W;	„ 8-A	(# 151)	„
NR-118W;	„ 8-A	(# 152)	„
X-119W;	„ 8-A	(# 153)	„
NR-115W;	„ 8-A	(# 166)	„
NC-167W;	„ 8-A	(# 167)	„

Serial # 140 later as 8 Special on floats with 575 Cyclone engine; serial # 143 later as Altair 8-D; serial # 144 also as 8-A; serial # 145 later as Altair 8-D; serial # 149 used in Mexico; serial # 150 was Sport Cabin Sirius and not on this ATC; serial # 152 later as Altair 8-D, became "Lady Southern Cross"; serial # 153 later as Altair, became Army YIC-25; serial # 167 later to 8-C as Sport Cabin Sirius.

From the connotations present in some of the accounts given here in this book, one could easily believe that the aircraft industry was really in dire straits and on the brink of folding up but actually this was not quite the case. True, many manufacturers were severely hurt and some were just on the verge of closing their doors but hardly none had lost faith or were diverted from their determination to keep developing the best airplanes that man could conceive; some of our finest airplanes were born in the next few years.

APPENDICES

BIBLIOGRAPHY

BOOKS:

Aircraft Year Book for 1930-1931
Wings For Life by Ruth Nichols
Jane's All The World's Aircraft by C. G. Grey
A Chronology of Michigan Aviation by Robert S. Ball
The Ford Story by Wm. T. Larkins
Revolution In The Sky by Richard Sanders Allen

PERIODICALS:

Flying   Western Flying
Aviation   American Modeller
The Pilot   Journal of A.A.H.S.
Aero Digest   Air Transportation
Air Progress   Model Airplane News
American Airman   Antique Airplane News

SPECIAL MATERIAL:

Licensed Aircraft Register by Aeronautical Chamber of Commerce of
America, Inc.
Characteristics Sheets by Curtiss Aeroplane & Motor Co.

CORRESPONDENCE WITH THE FOLLOWING INDIVIDUALS:

Chas. W. Meyers   Mark M. Campbell
Earl C. Reed   Howard W. Smeltzer
James W. Bott   Chas. F. Schultz
Peter M. Bowers   John R. Ellis
Douglas E. Anderson   Richard S. Allen
John W. Underwood   Edward J. Gardyan
Burton Kemp   Chas. E. Lebrecht

PHOTO CREDIT

Aeromarine-Klemm Corp. — Fig. 10.

Earl Adkisson — Fig. 126.

Douglas E. Anderson — Fig. 40.

H. W. Arnold — Figs. 12, 65.

Gerald Balzer — Figs. 1 (Mono Aircraft), 2 (Mono Aircraft), 3 (R. O. Bone Photo), 5 (R. O. Bone Photo), 56 (Mono Aircraft), 60 (Mono Aircraft), 61, 62 (Camera-grams), 63, 64, 70 (Fokker), 71 (Fokker), 73 (Fokker), 74 (Hughes Photo), 75, 80, 92 (General Airplanes), 93 (Hare Photo), 95, 105 (Reid Studio), 110, 111, 113, 124, 127, 131 (Waco Aircraft, 133 (Moreland Aircraft), 155, 158 (Art Streib), 159 (Mono Aircraft), 160, 170, 174 (Frank Griggs Photo), 180 (Waco Aircraft), 181 (Waco Aircraft), 187, 192 (Keystone Aircraft), 204, 207 (Paramount Air-craft), 209 (Paramount Aircraft), 225 (B. F. Parrish), 227, 237, 239, 244, 252 (Fokker), 253 (Fokker), 255 (Fokker), 256, 265 (Inman Photo), 277 (Kalec Inc.), 278 (Kalec Inc.), 280 (Detroit Aircraft), 283 (Spartan Aircraft Co.), 297, 300, 305, 308, 315.

Ball Studio — Figs. 53, 294, 309.

Beech Aircraft Corp. — Figs. 16, 245, 246.

Bellanca Aircraft Corp. — Figs. 146, 147.

Roger Besecker — Figs. 67, 167.

Boeing Airplane Co. — Figs. 20, 21, 22, 34, 35, 36, 37, 38.

Peter Bowers — Figs. 9, 11, 13, 14, 58, 59, 66, 76, 83, 97, 106, 117, 123, 199, 200, 205, 206, 224, 233, 235, 242, 270, 306.

Cessna Aircraft — Figs. 23, 24, 25, 26, 136, 137, 138, 139, 140, 141, 142, 143, 144.

Joe Christy — Fig. 33.

Consolidated Aircraft Corp. — Figs. 184 (Convair Div.), 186, 284, 285 (Convair Div.), 286 (Convair Div.).

Court Commercial Photo — Figs. 47, 128, 129, 130, 156, 272, 274.

Cresswell Photo — Figs. 189, 190, 191, 258, 292.

Curtiss Airplane & Motor Co. — Figs. 121, 222.

Davis Aircraft Corp. — Fig. 176.

Delta Air Lines — Fig. 194.

Harvey Doyle — Fig. 153.

Robert Esposito — Figs. 132, 197, 250.

Fairchild Aircraft — Figs. 46, 48, 134, 135, 282.

Fokker Aircraft Co. — Figs. 107, 254.

Ford Motor Co. — Fig. 303.

Hare Photo — Figs. 31, 32.

Stephen J. Hudek — Figs. 52, 72 (Fokker), 100, 102, 112 (Ford Motor Co.), 114 (Ford Motor Co.), 145 (Ford Motor Co.), 148 (Ford Motor Co.), 149 (Ford Motor Co.), 150 (Ford Motor Co.), 177 (Ford Motor Co.), 182, 201 (U.S.A.F.), 221 (Curtiss), 232, 249, 279 (Tom Towle), 301, 302, 304.

Marion Havelaar — Figs. 4, 49, 78, 183, 281.

Inman Photo — Fig. 266.

International News Service — Fig. 85.

Burton Kemp — Figs. 96, 104, 238, 251.

Keystone Aircraft — Figs. 193, 195.

Chas. E. Lebrecht — Figs. 291, 293.

Lockheed Aircraft Corp. — Figs. 81, 84, 169, 313, 314, 316.

Louis M. Lowry — Figs. 259, 262.

Macdonald Photo — Figs. 248, 264.

F. C. McVikar — Figs. 108, 228, 231.

Chas. W. Meyers — Figs. 86, 87, 88, 89, 90, 91.

Ralph Nortell — Figs. 6, 7.

National Aviation Museum, Ottawa, Canada — Figs. 157, 299.

Roy Oberg — Figs. 27, 101, 109, 125, 154, 196, 202, 208, 241, 307.

P & A Photo — Fig. 185.

Pan American World Airways System — Fig. 188.

Publishers Photo Service, Inc. — Fig. 312.

Earl C. Reed — Figs. 162, 236, 260, 263.

Schmidt-Bowers — Figs. 172, 311.

Smithsonian Institution, National Air Museum — Figs. 8, 15, 39, 41, 42, 43, 44, 45, 50, 51, 55, 68, 69, 77, 79, 82, 98, 99, 115, 116, 118, 119, 120, 122, 152, 171, 175, 198, 210, 211, 212, 219, 220, 223, 243, 257, 261, 267, 268, 271, 273, 275, 276, 295, 296, 298, 310.

Edgar B. Smith — Figs. 17, 247, 288.

Spartan Aircraft Co. — Fig. 269.

Stearman Aircraft Co. — Figs. 164, 165, 166, 168, 287.

St. Louis Aircraft — Fig. 234.

John W. Underwood — Figs. 54 (O. R. Phillips), 94, 151, 153, 178, 179, 203, 218, 229, 230.

United Air Lines — Figs. 18, 19.

U. S. Air Force — Fig. 217.

Alfred V. Vervile — Figs. 213, 214, 215, 216.

Truman Weaver — Figs. 57, 161, 163 (Clayton Folkerts).

Gordon S. Williams — Figs. 226, 290.

Williams-Schmidt-Bowers — Fig. 173.

Wolffoto — Figs. 28, 29, 30, 103.

INDEX